3

Contents

Return on or before the
last date stamped below

Learning Resources
Centre

For all human beings !

Forward

In this book we trace the development of Media and Cultural Theory from the Enlightenment through to the present day. Along the way we gesture towards a range of contemporary media texts including film, television, journalism, pop music, the Internet. And, indeed, what first attracted us to BookBoon was the opportunity to create a text that we could update and keep fresh; the media industry moves quickly and it is important to be able to revise interpretations in light of this. With this in mind we decided to make the scope of material covered fairly broad, ranging from Edmund Burke's notion of the sublime to Paul Gilroy's ideas about race and nationhood. Though roughly chronological, we have organised chapters by theme and ideas. Inevitably there is some crossover between the two and where possible this is gestured to in the flow of the text. However, it is our view that it is often very difficult to isolate specific strands of thought when discussing the complex ways in which media and cultural texts communicate.

The aim of the book then, is to provide students with an introduction to Media and Cultural Theory in a way that is both readable and engaging. So many media students shy away from theoretical application and this is a great shame as it is useful tool to illuminate and enlighten our understanding of the texts we consume. Though the core of the work is grounded in the delineation of key theoretical perspectives, the book is not trying to shed new light on any of the theorists discussed per se. Rather, it is our intention to explore how these theoretical perspectives might inform thinking about contemporary media and cultural production. In this direction, the book can be viewed as a starting point for students, guiding them as to how they might begin to incorporate the seemingly bewildering selection of theoretical perspectives on offer into their own work. Though we have endeavoured to provide the reader with useable précis of each writer's key work and useful bibliographical advice, it should be emphasised that there is no substitution for reading the original texts. In this sense we have endeavoured to write the book we always wished wish we'd had when we were students! And, for the visual learner Bevis has included illustrations! These are all original works and we feel they bring to life some of the more abstract concepts and ideas explored in each chapter. Further examples of his work can be viewed on his MySpace page.

The diverse range of theoretical perspectives covered in this book is of course a reflection of the varied nature of Media and Culture Theory, which draws upon aspects of sociology, linguistics, psychology, art-history and economics. It is also a reflection of the competing backgrounds of the authors! Stephen has completed a PhD in Cultural Studies, which focuses on the music press; he also teaches Media, English and Sociology. By contrast, Bevis has a Fine Art background and is completing a PhD in Human Geography; he currently teaches Graphic Art and Media theory. Indeed, it is our experience as students, researchers and teachers that has informed the shape and direction this book has taken. For our first collaboration we didn't want to produce a dry academic text, but rather a lively and engaging read that would hopefully provoke discussion and convey our own personal love of Media and Cultural Theory. We hope you enjoy it!

If you like to contact either of us then please feel free to do so through our MySpace pages:

Stephen Hill www.myspace.com/sah78uk

Bevis Fenner www.myspace.com/bijon77

1. Theories of the Enlightenment

Introduction

In this chapter we look at the ways in which three theories of the pre and post Enlightenment era can be to used to frame and shape the way in which we think about contemporary media, society and culture. The chapter begins with an overview of Edmund Burke's concept of the sublime and explores the way in which this can inform our understanding of contemporary media texts including film, pop music and news reporting. The second section of the chapter turns to focus on the work of Jeremy Bentham. In the final part of the chapter we fast-forward to the work of Mikhail Bakhtin in the Twentieth Century, whose writing invokes the pre-Enlightenment concept of the carnival, and consider the ways in which this chimes with postmodern cultural theory. First, however, we turn to what is actually meant by the term Enlightenment and the work of Immanuel Kant.

1.1 Immanuel Kant and the Enlightenment

NAME: Immanuel Kant (1724 – 1804)

KEY IDEA: That society can be bettered through the pursuit of understanding of the unknown through reasoned philosophical, scientific and aesthetic enquiry.

KEY TEXT: Answering the Question: What is Enlightenment? (1784).

The term Enlightenment is generally used to describe the period during the Seventeenth and Eighteenth Century during which Western society embraced rational thinking as a way of explaining both natural and cultural phenomenon. The rise of the Enlightenment is concurrent to the proliferation of many advances in medical sciences, the birth of the Industrial Revolution and other major changes in the way in which industrialised societies organised themselves.

Immanuel Kant's 'What is Enlightenment?' (1784) Is often cited as a definitive work. Kant views characterise the age of Enlightenment as 'man's emergence from his self-imposed immaturity':

> Immaturity is the inability to use one's understanding without guidance from another. This immaturity is self-imposed when its cause lies not in lack of understanding, but in lack of resolve and courage to use it without guidance from another. Sapere Aude! "Have courage to use your own understanding!" — that is the motto of Enlightenment (Kant, 1784, 1)

Central to this is the rejection of religious views and a view of society and culture that is ever evolving. In his logic, rationality and reason develops over time and systems of order and governance should reflect this. In particular he states that it is immoral for one generation to pass laws and doctrines that will inhibit the development of free thought in subsequent generations.

The measurement of things (Enlightenment)

If the technology that underpins contemporary media cultures has its origins in the industrial transformations that took place concurrent to the Enlightenment then in considering the way in which theory can be used to inform and shape our understanding of contemporary media practise we need to address that. Though cinema and recorded music may be a phenomenon of the Twentieth Century, other aspects of modern culture developed considerably during this period. For example, the mechanisation of the printing press revolutionised the production of literature. Likewise, the proliferation of the proscenium arch theatre made the fourth wall a dominant dramatic device in drama. However, it is perhaps the way in which society increasingly began to conceptualise entertainment and the arts as a popular and commoditised cultural form that is most significant.

1.2 Edmund Burke and the Sublime

NAME: Edmund Burke (1729 – 1797)

KEY IDEA: That the sublime is a distinct aesthetic category from the beautiful: a natural effect, which can often work in violent opposition to that which we perceive to be beautiful.

KEY TEXT: *A Philosophical Enquiry into the Origin of Our Ideas of the Sublime and Beautiful* (1757).

The awe of nature (the sublime)

Edmund Burke was a politician and philosopher born in Ireland in 1729. He was a prominent member of the original conservative faction of the Whig party. Burke was a strong critic of the French Revolution and his political philosophy is often seen as the forbearer of modern conservatism. In 1757 he published A Philosophical Enquiry into the Origin of Our Ideas of the Sublime and Beautiful a foundational work that was later taken up by Emmauel Kant in a philosophical exposition of the mechanics of aesthetic judgement (the ethics of taste) as part of his Critique of Judgment (1790).

Burke's work was groundbreaking in separating the sublime from the beautiful, the former of which was previously seen an aesthetic effect of the extremes of nature that worked in harmonious contrast with the latter. For Burke (1757), the sublime works in opposition to beauty, which is produced subjectively. He describes the sublime as being an external source of terror or 'whatever is in any sort terrible, or is conversant about terrible objects, or operates in a manner analogous to terror'. The sublime is also causal in the sense that while beauty can be seen as a subjective effect produced in response to a particular object, the sublime is induced by the object. In Burke's words 'the mind is so entirely filled with its object, that it cannot entertain any other, nor by consequence reason on that object which employs it' (Burke, 1757, II, 1).

The use of this single term eliminates the need for so many other adjectives that serve to distance the individual from pure experience. As Twitchell puts it, the sublime is 'an attempt at the farthest perceptual extreme to reconcile subject and object, self and nature' (Twitchell, 1983, 11). Despite this emphasis on objectivity and transcendence, the sublime is paradoxically a subjective contrivance based around individualistic notions of the self and its encounters in the world. In turn, it can also be seen as resistive of the dominant principles of the time - an attempt to break away from the subjective detachment of scientific reasoning and it's challenges to God. As Roberts suggests, 'when the gods withdraw from the world, then the world itself starts to appear as other, to reveal an imaginary depth which becomes meaningful in itself' (Roberts, 1994, 173).

Another way to understand the sublime is to deconstruct the term. The first part of the word 'sub' means below or up from underneath and the second part 'lime' comes from the Latin 'limen', meaning theshold or boundary. Taking this definiton into account what becomes apparent is that the state of sublimation is neither liminal nor transitional. In other words it is not an in-between state; instead only existing before or in anticipation of an action or event. It is a fictive moment that preempts an event which cannot actually occur – for in its occurrence the sublime would cease to exist and indeed could never have existed. It is this notion of detached proximity the gives the sublime its power of seduction.

Concurrent to the proliferation of Enlightenment thinking in the Eighteenth Century a parallel and quite contradictory aesthetic movement in the arts challenged rationale explanation; Romanticism emphasised the emotional origin of aesthetic experiences and the underlying element of fear in sublime experiences. Romantic landscape painters like J.M.W. Turner, for example, imbued their work with both expressive and disturbing qualities that caused controversy at the time. In Calais Pier (1803), for example, the straightforward appeal of the maritime scene is overshadowed by the awesome power of the sea. Likewise, Casper David Friedrich's The Wander above the Sea of Fog (1818) offsets the proprietorial gaze of the gentleman by a sense of the figure's insignificance in relation to the scale of the landscape.

This contrapuntal sensibility is of course a common convention in modern media. Think for example of the pleasure in watching a thriller film: the edge of your seat discomfort at watching those in peril. Alfred Hitchcock is of course the master of jeopardy and suspense. In The Birds (1963), for example, he invokes a sublime terror in the collective power of avian creatures. The romantic whim upon which Tippy Hedren purchases a pair of caged loved birds contrasts the dread of the choreographed aerial attack in the final scenes. Likewise, Steven Spielberg's Jaws (1975), which centres on a series of shark attacks at an American beach resort explores man's engrained fear of what lies beneath the ocean. Of course Spielberg's construction of the text emphasises the trepidation of the viewer: not least in its use of music. However, the key element of the film's success is the way in which it contrasts the light-hearted pleasure of the seaside with a primordial terror of natural elements.

Popular music is likewise peppered with examples of text that walk the fine line between pleasure and pain. Think of the number of songs that set lyrics of heartbreak and sadness to tunes that are uplifting and joyous. The songbook of Bjorn Ulvaeus and Benny Andersson (Abba), is exemplary of this. 'Knowing Me Knowing You' (1976), for example, tells the story of romantic separation and yet uplifting tune inspires pathos while the disco beat compels the listener to dance. Though 'Knowing Me Knowing You' might be moving, it does not perhaps embody the real terror inducing quality of the sublime. In this direction it is perhaps music that emanating from the punk era that best invokes the detached proximity of fear indicative of sublime aesthetic experiences. The music of the Sex Pistols for example embodies not only social alienation in terms of the lyrics but also musical alienation in the rejection of normative standards of instrumentation and playing. Of course, just as the music of the Sex Pistols adheres to fairly conventional song-structure in terms of composition, so too is the trepidation it induces restricted to the specific social context of punk. Far more sublime, in this sense, is the sonic terrorism of the British industrial group Throbbing Gristle. In the tradition of the musical avant-garde, the band's music rejects traditional song-structure and melody to provoke the listener into confronting their own expectations of what constitutes popular music. And, in this sense, it could be argued that the quartet, whose 'greatest hits' is entitled 'Entertainment Through Pain' is true to both Kant's conception of Enlightenment and Burke's take on the sublime.

11

Events in multi-media journalism are likewise characterised by the quixotic sensibility of the sublime. For example, from the 'Cold War' to the 'War on Terror' the nature and character of global relations is often characterised by an uncertain fear of an abstract threat. Though conflict is well documented in art, literature and film, it was perhaps the televisual spectacle of the Vietnam that brought the experience of war into the homes of millions of viewers at the end of the 1960s. The daily delivery of harrowing images from the front line generated exhaustion and revulsion that arguably culminated in America's eventual withdrawal.

That is not to say that television executives were as politically motivated as they were by the opportunity to secure viewers with footage that would shock and appal. However, without doubt, the contradiction between the visual depiction of large-scale human suffering and the comfortable distancing effect of the television could be said to be characteristic of the sublime. The same contradictory elements were of course in play for Live Aid in 1985, the benefit concert that took place to raise money and awareness for the famine in Ethiopia. Though the event was motivated by the altruistic sentiments of the Anglo-American music community, the spectacle of the suffering also served to infuse the music with new meaning. Most recently, the death of UK reality TV celebrity Jade Goody has embodied elements of the sublime. On the one hand, detachment from the star is emphasised by the way in which her physical decline was played out on the media stage. On the other hand, the very spectacle of her pain and suffering and the certainty of her impending death forced the audience to confront its own fear of mortality.

1.3 Jeremy Bentham, Utilitarianism and the Panopticon

> **NAME:** Jeremy Bentham (1748 – 1832)
>
> **KEY IDEA:** Utilitarianism: social utility as a measure of the overall happiness of a society. The utility principle is expressed as 'the greatest happiness of the greatest number'.
>
> **KEY TEXT:** *The Principles of Morals and Legislation* (1786).

Jeremy Bentham was a jurist, philosopher and social reformer born in England in 1748. He is attributed to the devising the principles of social utilitarianism. In his doctrine on legal jurisdiction The Principles of Morals and Legislation (1786), Bentham introduces the concept of social utility: a measure of the overall happiness of a society. The utility principle is expressed as 'the greatest happiness of the greatest number'; a phase Bentham borrowed from theologian Joseph Priestley's First Principles of Government (1768). The key remit of the code is to ensure the proliferation of pleasure and the negation of pain in society. He argues that '[n]ature has placed mankind under the governance of two sovereign masters, pain and pleasure. It is for them alone to point out what we ought to do, as well as to determine what we shall do' (Bentham, 1786, 2). Bentham is also concerned with the generalised way in which community is viewed as a homogenous body with the same interests and values, arguing instead that it 'is a fictitious body, composed of the individual persons' and therefore utility can be defined as 'the sum of the interests of the several members who compose it' (Bentham, 1786, 2). The term utilitarianism was later adopted by John Stuart Mill who qualified the pleasure principle with a series of qualitative distinctions as to the value of different types of pleasures to the individual and society as a whole.

Surveillance for your own good (Utilitarianism and the panopticon)

Many of Bentham's ideas have influenced our understanding of modern societies, in particular the relationship between the individual and the group. Key to this is Bentham's panopticon: a re-design of the prison building with the primary aim of improving its cost and efficiency. The design consisted of a circular building allowing for a single layer of tiered prison cells built around a central tower from which a single observer could view all of the cells. The cells were to be backlit and the central void surrounding the tower unlit so that observer could remain hidden whilst those in the cells would be constantly visible. The observation room would be darkened and it's windows obscured to disguise its occupant. The main function of the design was to reduce the number of employed guards by handing the task over to the prisoners themselves. The anonymity of those guarding, effectively meant that the prisoners could guard themselves, whilst the impossibility of telling whether or not the tower was occupied eliminated the need for continuous observation.

This situation would also serve to induce discipline as an auto-surveillance in which the observed would internalise the gaze of the observer and thus watch their self. It is this notion of self-observation - the watched doing the watching - that inspired Michel Foucault's notion of the panopticon gaze. Foucault argues that as with the panopticon, discipline and the mechanisms of power and social control are embedded within institutions like schools, hospitals, and factories and, on a generalised level, within public consciousness. He suggests that the disciplinary model of power has 'infiltrated' other forms 'serving as an intermediary between them, linking them all together, extending them and above all making it possible to bring the effects of power to the most minute and distant elements' (Foucault, 1977, 478). In other words, diffusing the vernacular of discipline amongst all forms of social power from institutions, to groups, right down to the individual striving for autonomy ensures increased economic and political efficiency, effectiveness of power and growth in the 'output' of the institutions 'within which [power] is exercised; in short to increase both the docility and the utility of all elements within the system' (Foucault, 1977, 479). Here the idea is that discipline is so embedded at every level that we have all become compliant with and complicit in the power systems and happy to sing from the same song sheet, so to speak. Key to the complicity of individuals is their internalisation of the 'apparatus' of power: We can argue that our reliance on knowledge of institutions of power (including the media) or what Giddens (1990) refers to as 'expert systems', the self-reflexivity (self-scrutinising) that enables us to construct and perform our self-identities, and the self-surveillance required in observing social norms and conventions in what Goffman (1959) calls 'front-stage' performances, - those for the benefit of others - are all internalised forms of external power.

At a most basic level, it is possible to see Jeremy Bentham's notion of utilitarianism as the premise on which broadcast media is based. For example, the Royal Charter, which governs the role of the BBC as a public service broadcaster stipulates that the corporation must sustain citizenship, promote education and learning, stimulate creativity and cultural excellence. In effect, the remit of the BBC is to bring the greatest good to the greatest number. In addition to this, the charter legislates for the role played by the BBC in mediating relations between Britain and other sovereign states. For example, it states that the BBC should bring the UK to the world and the world to the UK. Likewise, the most recent charter (2007) states that the corporation should deliver to the public the benefits of emerging new media technologies.

Of course, there is a difference between what is in the interest of the public and what the public wants as the distinction between the BBC and commercial television is testimony to: since its inception in the mid-1950s independent television has tended to be viewed as more low-brow in its program content; appealing to the less cerebral impulses of television audiences. This hierarchy of pleasures is therefore commensurate with John Stuart Mill's conception of the term. That is not to say, however, that all commoditised cultural forms are incompatible with Bentham's model of utilitarianism.

During the 1980s the Conservative government in the UK embraced a series of reforms to the public sector that reflected the rise of electoral engagement with the role of the individual as consumer, as opposed to producer. As a feature of the generalisable proliferation of postmodern culture, the decline of heavy industry and domination of society by information technology, Tory reforms pertained to be at the vanguard of social utilitarianism. Indeed, central to Margaret Thatcher's policy of governance was financial expedience in the public sector. By appropriating the principals of free-market capitalism, Thatcher's reforms encouraged the electorate to engage as consumers of public sector services. This included the publication of school league tables, the privatisation of nationalised transport systems and the sub-contraction of auxiliary services in the NHS. The short-term effect of this was that the Conservative government was able to lower taxes creating considerable 'happiness' for middle-income earners. The long-term corollary, however, was that public sector services suffered not only from chronic under investment but that the selfish interests of stakeholders undermined the premise of egalitarianism in the provision of state services.

Other examples of the misappropriation of social utilitarianism include CCTV and speed cameras. While both of these phenomena are designed to protect the interests of the majority from the anti-social behaviour of the minority, the aggressive use of these technologies has in many instances had the reverse effect. Indeed, it could be argued that the proliferation of close circuit television is a very contemporary example of Bentham's panopticon: residents know it is not possible for somebody to be watching each of the cameras, however, their very presence modifies behaviour. Likewise, speed cameras offer a punitive incentive not to contravene road traffic laws, however, their conspicuous presence is a continual reminder that our actions are being monitored. Of course the very notion of Big Brother has its origins in George Orwell's dystopian vision of the future in Nineteen Eighty Four (1948): CCTV in this instance predicted by the two-way telescreen. And, indeed, the popularisation of this concept in the reality television program Big Brother (Channel 4) offers an ironic commentary on this ugly aspect of contemporary society. Its stars are able to subvert the subjudicating function of surveillance by objectifying the self for the cameras.

Most recently, however, it is the proliferation of social networking sites like Facebook, Twitter and Bebo that have lured us into a more intimate panopticon. In theory, these sites serve the very utilitarian function of allowing us to communicate more effectively with loved ones and friends. Increasingly, however, the awareness that other people are sifting though our photos and 'status updates' induces a self-reflexive panopticon gaze of its own. As the ubiquitous mobile phone camera is testimony: in the Twenty-First Century, increasingly our social existence is but a prop to support the version of our life we project into cyberspace. Indeed seeing and being seen is a primary focus of youth culture. As Hebdige suggests subculture 'forms up the space between surveillance and the invasion of surveillance, it translates the fact of being under scrutiny into the pleasure of being watched. It is hiding in the light' (Hebdige, 1988, 35).

1.4 Mikhail Bakhtin and the Carnivalesque

NAME: Mikhail Bakhtin (1895 – 1975)

KEY IDEA: The Carnivalesque: social and aesthetic formations, which disrupt the normative behaviours and socio-cultural hierarchies of everyday existence. Central to this is the idea that normal life is suspended during the carnival.

KEY TEXT: *The Principles of Morals and Legislation* (1786)

Mikhail Bakhtin is a Russian philosopher who was born in 1875. He is often associated with Russian Formalism and was an influence upon neo Marxist thinkers. His most influential work is a dissertation he wrote during the Second World War entitled Rabelais and His World. The thesis, which focused on the work of the French Renaissance writer Francois Rabelais was not published until 1965. In particular his work is said to have influenced Jacques Derrida and Michel Foucault.

In Rabelais and His World Bakhtin offers fresh analysis of the five-volume novel The Life of Gargantua and of Pantagruel that Rabelais wrote in the Sixteenth Century. Bakhtin argues that the text has been misunderstood and that certain sections have been repressed. His analysis, which reconfigures the text within the context of the Renaissance social system emphasises the significance of the carnival and grotesque realism. In particular Bakhtin's use of the term carnivalesque has been particularly influential. He repositions the term so that it refers not simply to the festival traditions of Northern Europe but encompasses the semiotic conventions of literary works like that of Rabelais.

Life turned inside out (the carnivalesque)

Though Bakhtin attributes the origins of the carnivalesque to the Feast of Fools, he contends that the term can be applied to aesthetic forms, which disrupt the boundaries of everyday existence. Central to this is the idea that normal life is suspended during the carnival. In particular it subverts traditional hierarchical structures and forms of 'terror, reverence, piety and etiquette connected with it' (Bakhtin, 1965, 250). For Bakhtin a key component to the carnivalesque is that all distance between people is suspended and a 'special carnival category goes into effect: free and familiar' (Bakhtin, 1965, 250).

The influence of Bakhtin's work and the carnivalesque can be found on a number of contemporary writers and thinkers. For example, the notion that there is a semiotic space separate to the mainstream (non-carnival) is pivotal in sub-cultural theory. In Subculture and the Meaning of Style (1979) Dick Hebdige talks about the way in which audiences rework the meaning of signs according to the social context of their use with specific reference to punk. Likewise, in Postmodernism or the Cultural Logic of Late Capitalism (1991) Frederic Jameson explores the way in which the satiric impulse of parody has a semiotic intent that distinguishes it from the more neutral goal of pastiche.

Though the origins of the term carnivalesque can be traced back to the customs of Renaissance society, it continues to be very relevant to the study of contemporary media. In part, this can be attributed to the fact that semiotics and Russian Formalism influenced Bakhtin's work. However, much of his acclaim was posthumous: and it was Tzvetan Todorov and Julia Kristeva who bought his work to the attention of the French speaking academic world after his death in 1975. The resurgence in interest in Bakhtin's work on the carnivalesque can be attributed to its compatibility with contemporary definitions of postmodern culture. In particular its emphasis on the interplay between appearance and reality echoes the way in which Baudrillard and Jameson contend that postmodern society is characterised by the collapse of the distinction between the real and the simulated.

Of course this is in part a function of the proliferation of information technology: from the advent of the telephone to the internet, the mediation of human interaction by machine has blurred the boundaries between authentic communication and that which is virtual. Moreover, the routinised way in which such technologies have become embedded in all our lives means that that distinction is not always particularly significant. Who is to say whether a conversation that takes place on the telephone is less genuine than a dialogue that takes place face to face?

The usefulness of the carnivalesque then, is that it takes that plurality as its starting point. As Bakhtin suggests, normal life is suspended and traditional hierarchies subverted. In this sense, the preference for social interaction that embraces technology is carnivalesque because it challenges received ideas about the authenticity of face-to-face communication. If we extend this thinking beyond the realm of human interaction to explore the ways in which people in the Western World spend their leisure time it is easy to see other facets of contemporary media consumption as carnivalesque.

In this sense it could be argued that the willing suspension of disbelief endemic to the cinema audience's sense of engagement with the text is characteristic of the carnival. The abandonment of everyday reality is for example central to the carnival-like pleasure of going to the cinema. And, indeed there is a sense in which the icons of stage and screen embody carnivalesque ideas about the fool king. However, what this misses is the underlying self-consciousness that characterises carnivalesque actions and their ironic reach. Engagement with a postmodern cultural form is not enough, but rather it is the audiences' awareness of the artificiality of that experience that transgresses normative semantic fields.

In contemporary culture the carnivalesque is often associated with cultural matter that might be labelled 'camp'. Indeed, there is considerable mileage in the way in which Susan Sontag uses the term to describe a Foucauldian semiotic event. Bakhtin's work is less concerned, however, with gender and for this reason as we move towards a society less pre-occupied with narratives of patriarchal oppression, the carnivalesque is a useful term to describe the self-conscious and playful way in which media consumers engage with contemporary media texts.

Modern cinema is littered with examples of the carnivalesque: films that not only play with the audiences' perception of the text but also incorporate that ambiguity in the way in which meaning is constructed. Stanley Kubrick's A Clockwork Orange (1971) is a heavily stylised film in which choreographed dance routines are used to offset ritualised violence. The effect on the audience is mimetic: forcing to them to confront the complicity of their own pleasure in consuming the sadistic and cruel. While Kubrick's masterpiece is often considered an exercise in good film making, John Waters Desperate Living (1978) inverts the normative conventions of cinema by celebrating the trashy and distasteful. Dialogue is wooden, sets shoddy and acting is over the top in Waters' low budget production, which purposefully subverts the semantics of Hollywood cinema. Furthermore the film involves a carnivalesque world (Mortville), which is ruled by a false queen who instigates inverse events. More mainstream examples of the carnivalesque can be found in the work of Baz Luhrmann. Romeo + Juliet (1996) is set in the dystopic vision of a near future Hollywood and is cohered around the debauched revelry of a masque ball: the way in which Luhrmann augments the script with contemporary popular songs and dance routines inverts the reverence normally bestowed upon Shakespearean texts. Likewise, though Moulin Rouge (2001) is set in Paris at the end of the Nineteenth Century, the original soundtrack includes songs by Elton John, David Bowie, Christina Aguilera and Fat Boy Slim.

Baz Luhrmann's use of the popular song to challenge audience expectation is characteristic of the way that pop music culture embodies many aspects of the carnivalesque. For example, the parodic sensibility of the pop scene in Britain during the 1950s was extremely carnivalesque in its appropriation of the more of mainstream Afro-American pop culture. Stars like Cliff Richard and Adam Faith were, for example, really facsimiles of Elvis and other American stars. Likewise, chameleon-like figures like David Bowie, Annie Lennox and Madonna blur the boundaries between the real the simulated: the authentic and the fake. It is not surprising therefore that the arena in which popular music culture is most carnivalesque is the music video. Often described as the ultimate example of postmodern practise, pop music videos (which at their most authentic depict musician miming to synchronised backing tracks), embody a free-form world in which every flavour is on the menu and no style seems to clash. A contemporary example of this is Denis Thibaut's production for Bob Sinclar's 'Rock This Party' (2006) in which the fifteen year old actor David Beaudoin adopts the identity of whole host of stars from the history of popular music including Michael Jackson, Nirvana, AC/DC, Bob Marley and Justin Timberlake. The way in which this is then edited to different parts of the track (itself a sample based composition) reinforces the pantomime like quality of the visuals. More so than perhaps any other aspect of contemporary culture, the fluid and celebratory nature of modern pop culture embodies the pre-Enlightenment sensibility of the carnival.

Conclusion

In reviewing three theories of the pre and post Enlightenment, it is easy to see how they sit as cornerstones of modern thinking and can offer fascinating insights into the way in which contemporary texts make meaning. Most interesting perhaps is the tension that exists between the romantic visions of the sublime rooted in emotional resonance and authenticity, and those formations influenced by rational thinking and reason. In part this is because this debate is the backdrop to contemporary ideas about romanticism and modernity. Likewise, the fine line between utilitarian function (pleasing most of the people most of the time) and the more repressive impulses of the panopticon gaze have direct relevance in a society dominated by information technology. Ironically however, it is perhaps the pre-Enlightenment concept of the carnival that has most to offer contemporary thinking because it encapsulates the wilful subversion and playfulness of a culture long since enveloped in a very postmodern sensibility.

2. Marxism and Global Media

Introduction

In this chapter, we look at how classical Marxism can inform our understanding of the political and economic relationships underpinning global media. We argue that although the Communist Revolution may have failed to materialise in the UK, transformations in global media raise interesting questions about how we interpret Marx's work. In particular, the chapter focuses on the international nature of the relationship between the Proletariat (the workers) and the Bourgeoisie (the ruling class) and argues that access to the ownership of the means of cultural production as opposed to material production that is definitional of political power. To revisit the work of Karl Marx is perhaps pertinent to the current shift in world politics brought about by America's appointment of Barack Obama: marking a departure from years of American centre-right politics. Moreover, the reliance of his campaign on Internet generated funding highlights the way in which new media technologies have redefined the political landscape of the Twenty-First Century. First, however, we begin with an over-view of the work of Karl Marx and Frederic Engels on the nature of capitalism in The Communist Manifesto (1848).

Capitalism

While a Communist revolution may have failed to materialise in the West during the Twentieth Century in the way that Karl Marx predicted, the manifesto he wrote with Frederick Engels in 1848 remains the definitive account of the advent of capitalism. The document, which was originally published as a political pamphlet, plots the social consequences incurred by the economic changes of the Industrial Revolution at the beginning of the Nineteenth Century. Though the term capitalism precedes the Industrial Revolution, and is essentially a refinement of the economic liberalism Adam Smith talks about in The Wealth of Nations (1776), the changes to agriculture and industry in the UK bought about by mechanised communal production emphasised its underlying principals. For Marx the key idea is that during this period society moved from a body of self-sufficient private producers to an isolated mass of workers with no rights to the produce they make.

2.1 Karl Marx and Frederic Engels - The Communist Manifesto

NAME: Karl Marx (1818 to 1883)

KEY IDEA: The economy is at the base of society; everything else is determined by it. Under capitalism, the economy is exploitative: serving only the interests of the ruling class (the Bourgeoisie). This inequality will lead to revolution, which will be characterised by the workers (the Proletariat) seizing control the means of production and the end of capitalist economic exchange.

KEY TEXT: *The Communist Manifesto* (1848).

Marx believed, like the German philosopher Georg Hegel, that history moves dialectically: 'Nature works metaphysically; she does not move in the eternal oneness of a perpetually recurring cycle, but through a real historical evolution'. In any period in history Hegel believed that a generally accepted set of ideas can explain society called the thesis (e.g. capitalism). A set of alternative ideas then develops in response to this called the anti-thesis (e.g. socialism). Ultimately a hybrid of the two evolves, creating a new thesis and a new chapter in history (e.g. neo liberalism). However, Marx and Engels did not agree with Hegel that man's ideas shape society but that society shapes man. Thus Marx's and Engel's view of the dialectic differs from Hegel in that each stage in history is not determined by a new set of ideas but rather by a new way in which society is organised. This is called 'dialectic materialism'.

NAME: Frederic Engels (1820 to 1895)

KEY IDEA: Influenced by Hegel and Heraclitus, Engels contribution to the *Communist Manifesto* is that of 'Dialectic Materialism'. Change in the economic structure of society works through the dialectic principles of conflict between thesis and antithesis. In his logic the emergence of a synthesis of the two, i.e. a new economic thesis is characteristic of a new phase in history.

KEY TEXT: *The Communist Manifesto* (1848).

Marx and Engels believed that historically where conflict had arisen over the way in which society was organised it usually had a lot to do with disputes about the ownership of the means of production and inequalities created by the economic system. They consider the economic system to be the foundation of society underpinning judicial and political institutions and explaining the religious and philosophical ideas of any given historical period. For Marx the conflict between people is simply a reflection of the contradictions of the economic system. In his logic the ideal state of human nature is 'natural man': self-sufficient and free from exploitation. Capitalism for him is intrinsically corrupt in that man is alienated from the means of production and use value is secondary to that of exchange. Indeed, 'natural man' is a simple creature with basic requirements. Man spends all of his time in production of these – clothing, food, water, shelter etc – and is in harmony with nature.

This utopian vision corresponds quite clearly with Marx's view of the ages before capitalism. Production is on a small scale and private, as is ownership. Any exchange or trade is on a fair basis. If it takes one man two hours to catch 20 fish and another man about two hours to make one coat then one coat is worth 20 fish. Capitalism is born when the fisherman devises a way of catching forty fish in the time it takes to make a coat. In a pre-capitalist age this would mean that the coat was now worth forty fish. However, instead of paying the exchange value for the coat, the fisherman carries on paying twenty, which is the commodity value. The remaining 20 fish that the fisherman has caught that are not spent on the coat become 'surplus value' or profit.

Profit can then be spent on other things. However, the key point for Marx is that in order for fisherman to make a profit then the tailor must be exploited. Being a capitalist, the most probable thing the fisherman will spend his profit on is labour. By paying a labourer to help him transport, store and clean the fish he catches the fisherman can now optimise his productivity. Where he used to able to catch 40 fish in the time it takes the tailor to make a coat, he can now catch 80 with the help of wage labour. Though he pays the wage labour the equivalent of 20 fish for his trouble, with a coat still costing 20 fish the fisherman's profit is up 100% to 40 surplus fish. And, because communal production is so much more efficient and economical than private production it catches on: wage labour becomes the dominant mode of production. However, while production is social, ownership remains private and for Marx - being a dialectical materialist - this fundamental transformation in the economic base of society is indicative of a shift in the way that society is organised. Now being a clever and careful capitalist, the fisherman soon realises that he can use some of his profit to buy wood and to pay the carpenter to build him a boat. This has lots of benefit (for the fisherman): lots more fish and lots more profit! And so the process continues ad infinitum with more boats, more workers and more fish, until some financial mismanagement up-sets the apple cart. However, it also has a number of key disadvantages.

Return to nature? (The Communist Manifesto)

The contradiction between ownership, which is private, and production, which is social, creates animosity between the 'owners of the means' of production and 'the workers'. The division that now exists between these two groups is characterised in the terms Marx uses to describe them: the Bourgeoisie and the Proletariat. Just as the tailor has to be exploited in order for the fisherman to create surplus value, so too does the Proletariat have to be exploited so that the Bourgeoisie can make a profit. Consequently capitalism can be said to contain a number of inherent contradictions:

1. As wages are kept to a minimum to ensure maximum profit, capitalism simultaneously destroys its own market.

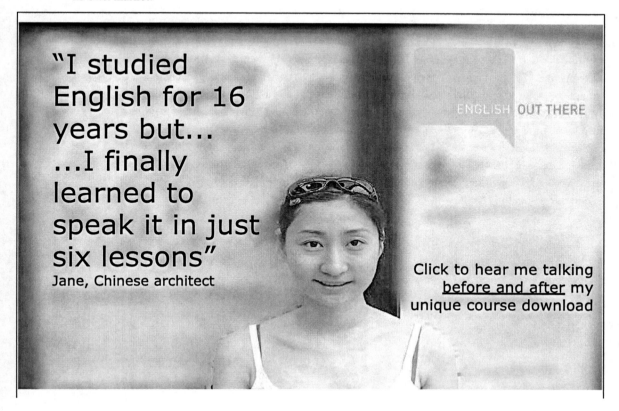

2. People cannot afford goods and so supply becomes greater than demand.

3. Consequently weaker capitalists go out of business and have to join the ranks of the Proletariat and become workers.

Marx argues that these contradictions could be resolved in a number of ways. Firstly he suggests trusts could be organised with agreed prices and production methods. Secondly, he suggests that state could interfere and take over the 'direction of production' and ownership of its means. However, Marx argues that neither of these methods is likely to work because:

1. Capitalists in trusts would simply cheat and produce more than has been agreed.

2. The state would run the means of production as capitalists it would, therefore, still embody all the contradictions of capitalism

Consequently Marx believes the only real solution is Communist Revolution: this involved the Proletariat seizing political power and turning the means of production into state property. The Proletariat then re-arrange the state into a communist operation and at this point the state itself dies out: for Marx the state becomes unnecessary when it is truly representative of society.

What Revolution?

Of course, in Britain, as with most of the Western capitalist economies, the revolution has failed to materialise, which raises questions about the usefulness of Marx's predictions. And, in countries where revolution has taken place (like Cuba and Russia), this has been characterised by strong structures of state, which hardly resemble the utopian vision held by Marx. So, 160 years on from its publication, it is easy to reject the blueprint for social change outlined by Marx and Engels in the Communist Manifesto. Perhaps part of the problem is that Marx's vision of economic activity is limited to the political landscape of an individual sovereign state: it did not anticipate the complex political and trading alliance that characterise contemporary global relations. This is perhaps surprising given that the period in which Marx was writing is often described by historians as the Imperial Century: a period in which 10 million square miles were added to the British Empire. In this sense the Proletariat/Bourgeoisie model is a less than exact fit when it comes to the arrangement of domestic labour in the Nineteenth Century. Historically Britain's wealth was based on Empire and owed as much to the exploitation of the colonies as it did UK based workers. This will be discussed in more detail in chapter 10.

This international dimension to capitalist economic relations has come to the fore in recent years with the decline of both heavy industry and manufacturing in many First World economies. In the UK, the skilled labour force of working class cities like Sheffield, Glasgow and Manchester have been unable to compete with the lower wages and minimum employment legislation in the Developing World. Such is the global nature of multi-national organisations that in Marxist terms it could be said that by and large the West now out-sources for Proletariat (workers). Companies like Nestle and MacDonalds, for example, ensure maximum profitability by operating wherever possible in Third World countries: where labour is cheap and capitalism is unimpeded by concerns for human rights of the environment. In this sense the relations and contradictions of economic exchange outlined by Marx in the Communist Manifesto fit our global economy very well. The developed world stands comfortably in the shoes of the Bourgeoisie, with the developing world assuming the subordinate position of the Proletariat. This raises the question, are we premature then in thinking that the revolution has failed to materialise? To answer this, it is perhaps necessary to think about what it means to be a member of the Bourgeoisie (the ruling class) in the Twenty First Century.

2.2 The New Bourgeoisie

In Twenty-First Century, many people in Britain could be defined as bourgeois without necessarily being 'owners of the means of production'. In part, this can be ascribed to the legacy of compulsory education in the Twentieth Century, the Grammar School system and the emergence of a knowledge-based economy. As Pierre Bourdieu claims in Distinction (1979), education is at the heart of social class and consequently a more educated society is inevitably a more middle class one. However, the upward mobility of the Western lifestyles has not happened in isolation and can be seen as a direct consequence of the exploitation of other parts of the globe: from the tea plantations in India to the decimation of the Brazilian rainforests. To live in a Western economy is, therefore, to acknowledge our relative advantage over rest of the world's population: not just economically but in terms of life expectancy, health care, access to clean drinking water etc. The kind of poverty that underpinned the Industrial Revolution in England in Marx's lifetime is unthinkable today. It is perhaps unsurprising, therefore, that when former Deputy Prime Minister John Prescott asked a group of un-employed teenagers what class they thought they were in a BBC documentary on social hierarchy in the UK they replied most assuredly that they were 'middle class'. Now of course their use of the term cannot be understood in terms of old style definitions of relative social ranking: not least because the girls had neither property nor education to mark them above the ranks of the working class. However, in global terms their relative position is very much that of the Bourgeoisie. Therefore, how we begin to define the bourgeois requires careful consideration.

The decline of heavy industry

Ownership of the means of Cultural Production

To understand how to define the ruling class (Bourgeoisie) in a global society it is perhaps useful to return to the time of Marx's writing: the Imperial Age of the Nineteenth Century and the rapid expansion of the British Empire. When considering this achievement the question that hangs in the air is how did Britain, a small island with a modest population, achieve economic supremacy over half the globe? Clearly, our military might, underpinned by our maritime experience as an island nation, played a part in this, as did out technological advancement courtesy of the engineering giants like Isambard Kingdom Brunnel. However, what also made the British Empire possible was the communication system: the All Red Line, as it was known, enabled messages to be sent from Ireland to Newfoundland and from Suez to Bombay, Madras, Penang, Singapore and Australia. Rudimentary and undeveloped though the telegraph was, it put in place a nervous system that co-ordinated the export of British models of education, democracy and religion to the four corners of the globe.

Though Brunnel's ship, The Great Eastern, laid the first transatlantic cable within Marx's own lifetime, ultimately it is The Communist Manifesto's inability to foresee the way in which ownership of the means of cultural production would become definitional of the ruling class in the Twentieth Century that limits its relevance today. The last 100 years has seen some very real global conflicts over territory and space. However, in the intermittent periods of peace it often access and control of cultural production that has shaped the path of modern history. From Nazi propaganda to the censorship of mainstream Hollywood cinema in The People's Republic of China, global tension is played-out in the political relations underpinning the media economy. Testimony to this is the Cold War between Soviet and American government in the second half of the century: an ideological stalemate that brought the threat of nuclear Armageddon to the forefront of public consciousness, in everything from the race to put man on the moon to popular fiction about espionage and military coalitions. Though the US's emergence as the world's biggest super power can be explained by its role in the Second World War, as well as conflicts in the Middle East and South East Asia, underpinning this has been the not so silent march of Western culture in the form of entertainment, food and clothing. In this sense, we are all owners of the means cultural production because it is our culture that dominates the globe.

2.3 Cultural Imperialism and the Media Revolution

Though Britain has long since handed back its empire, we live today in an age of Western Cultural Imperialism that has its origins in the British Empire and the Americanisation of much of the Developed World in the Twentieth Century. Few major cities in the world are without access to a MacDonald's or a Starbucks and their aggressive economic principals exemplify all that Marx feared about the capitalist system. By the same token, the increasing individualism of audience engagement with media forms experienced through digital technology and Internet gives people in the West a far greater grip on culture than those in developing nations. That many of those people covert Nike sportswear and learn to speak English from American films reinforces this commanding position. However, since the terrorist attacks of 9/11, when hi-jacked airliners were flown into the twin towers of the World Trade Centre and the Pentagon in Washington, global politics has become increasingly characterised by headline grabbing media stunts, as opposed to traditional channels of political expression.

Ou est le weapons of mass destruction?

Exemplary of the self-conscious nature of political protest in the Twenty-First Century was an incident in the Russian province of North Ossetia in 2004. When Chechen rebels took the 1100 pupils and staff hostage at the Beslan School, before massacring 334 of them, digital video cameras were positioned allowing footage to be broadcast on the Internet. Likewise, the capture and murder of British and American hostages by Islamic extremists in Iraq has been highly sophisticated in its manipulation of International media. Most notable were the murders of Americans Eugene Armstrong and Jack Hensley along with British citizen Ken Bigley: a story, which dominated headlines for weeks after their capture on September 16th 2004. At various intervals, videos of the captives reading out political statements addressed to their respective governments and begging for assistance were posted on fundamentalist web sites and subsequently picked up by international news agencies. The media circus, which preceded their eventual beheadings on September 20th, 21st and October 7th respectively did little to save them and played right into the hands of the extremist groups: making anti-Western hostility the centre of public debate and highlighting the chaos ensuing from the Allied invasion of Iraq in 2003.

That DVDs of the beheading of American soldiers retail in Baghdad for 75 pence, while the execution of Saddam Hussein has over 2 million hits on YouTube makes it clear that technological innovation is such that the revolution will not be characterised by a call to arms, but that it is happening all around us in the media. From Al Qaeda to Fathers for Justice, headline grabbing media stunts are now at the forefront of world politics. We are living in an age in which political expression is becoming more ephemeral as the traditional channels of expression seem less tangible. Not only do the party politics of Westminster seem redundant in the face of global fear and the "war on terror" but also domestic issues of education, welfare and health seem to slide further from the prevailing climate of expectation.

Conclusion

Marx argued that it is the consciousness of man that determines society. On that basis, man has the power to be the agent of social change providing that change can be imagined. Increasingly, however, it would seem that it is our media consciousness that determines our social position within the global economy. The predominance of information-based industry requires cultural knowledge and not just financial resources. Moreover, the proliferation of information technology in the Twenty-First Century has the potential to undermine international relations of state and power that have endured since the days of the British Empire. The rise of China's economic importance and the Tiger economies of South East Asia (Hong Kong, Taiwan, Singapore and South Korea) highlight this. Most recently, the election by the American people of Democrat candidate Barack Obama on November 4th 2008 is perhaps a reflection of a change in Western perceptions of the changing nature of its own global position.

3. Semiotics and Formalism

Introduction

In this chapter we look at the way in which three semiotic theories can be used to frame and shape the way we think about contemporary media, society and culture. The chapter begins with an overview of Ferdinand de Saussure's model of the sign system and explores the arbitrary relationship between the signifer and signifed in the composition of media signs. The second section of the chapter turns to focus on the work of Valentin Volosinov and the concept of multi-accentuality: the way in which the meaning of a sign changes over time and according to context. Finally, in the concluding part of the chapter we turn to the work of Vladimir Propp and Tzvetan Todorov in a consideration of the formal structures of text level meaning including characterisation, narrative and genre. First, however, we consider what is actually meant by the term semiotics and the work Charles Sanders Peirce.

3.1 Charles Sanders Peirce and Semiotics

The term semiotics came into use towards the end of the Nineteenth Century to describe the logic of philosophy. Though it is often viewed as synonymous with the work of the Swiss linguist Ferdinand de Saussure, it was in fact the American Logician Charles Sanders Peirce who first coined the term. Peirce's work was not confined to abstract thought, however and the systems under-pinning his early experimentation with electrical circuits informed developments in computer technology. Reflecting his wider interest in Mathematics and Philosophy, Peirce's view of semiotics was more nebulous. In 'Logic as Semiotic', for example, Peirce describes the semiotic as the 'formal doctrine of signs (Peirce, 1940, 98). However, it his definition of what a sign is that is perhaps most useful:

> A sign, or representation, is something which stands to somebody for something in some respect or capacity. It addresses somebody, that is, creates in the mind of that person an equivalent sign or perhaps a more developed sign. (Peirce, 1966, 99)

Clearly language is the most logical exemplification of this: a word symbolising an object, action or subjective attachment. The noun 'tree', for example, can be customised by the adjective 'beautiful' or the verb 'run' by the adverb 'determinedly'. And indeed, in this sense, Peirce groups signs into two categories: 'representamen' (objects) and 'interpretants' (modifiers) (Peirce, 99). However, where Peirce's work is transforming is the invocation of this in relation to non-linguistic matter. For example, Peirce considers the grammar of visual signs including photographs, technical drawing and the animal world. Particularly instructive is his consideration of the zebra: its resemblance to a donkey he argues invokes an expectation that the creature will be obstinate. In this sense the meaning of the zebra is being modified: interpreted through the more familiar lens of our experience with donkeys. In a more contemporary context, therefore, this same process can be observed in our consumption of media texts. Our familiarity with Western genres, for example, is often used as the key to unlocking meaning in texts from other parts of the world. The use of the term Bollywood to describe films produced in the Indian city of Mubai (formally known as Bombay), for example, draws upon Western familiarity with the term Hollywood to denote the American film industry. And, as Edward Said notes in Orientalism (1978), such practices that fetishise the 'otherness' of non-Western cultural forms, actually serves to re-enforce the hegemony of the West. The politics of semiotics are, in this sense, a complex arena: issues of representation are of course central to studying media.

3.2 Ferdinand de Saussure and the Cours de Linguistique

NAME: Ferdinand de Saussure (1857 to 1913)

KEY IDEA: Structuralist approach to semiology; Saussure argues that all signs are double entities made up of the signifier and the signified. The signifier is the linguistic coding of a concrete object, abstract emotion or physical act. The signified is that to which the signifier refers to. The two things are inseparable; however, the relationship is arbitrary: meaning that there is no causal reason why the two are so related. The fluctuation of meaning in the relationship between sound and meaning across different languages is testimony to this fact.

KEY TEXT: *Cours de Linguistique Générale* (1916)

Drawing upon the work of Peirce, in Saussure's posthumous Cours de Linguistique (1916) the Swiss theorist also argues that the sign is the basic unit of exchange. This he suggests is composed of two parts the signifier and the signified. The signified is the object, condition or circumstance that is being referred to, while the signifier is the way in which that is culturally coded. For example, a perennial woody plant may be coded linguistically by the word tree: the correlation between an arrangement of letters and a phonetic sound. The signifier is the tree and the signified is the woody plant. And, the same can be said about a whole range of phenomena in the natural world and its cultural codification by humans.

There is, however, even greater fluidity when we think of abstract nouns: words like love, hate and jealousy. The relationship between these words and the state to which they refer is much more difficult to fix. Further complexity is added to the signs system by the variation of terms across different languages. For example, a tree in German is denoted by the word baum, while in French it is known as arbre. For this reason Saussure argues that the relationship between the signifier and the signified is arbitrary: determined by chance whim or impulse; though it can be observed across languages that some words have certain phonetic qualities in common that link with what they signify. There is, for example, often a gradation of the positive, comparative and superlative as evidence that language is not a totally arbitrary system: blank, blanker, blankest; the more phonemes, the more emphasis. That said, there is general consensus that in its basic form the sign is subjective. Having established this, Saussure then examines the ways in which signs are put together to make meaningful combinations of words.

The arbitrary relationship between the signifier and the signified

On studying the development of language Saussure asserts that there is no point studying its historical development (the diachronic perspective) as language can only be understood in the context of the instance of its usage (the synchronic). For Saussure, the synchronic context is all. And, the definition of something synchronically is as dependent on what it is not as what it actually is. For example, the definition of a raised surface to sit on as a chair is as dependent on what is not a chair as the reality of the chair itself. In this sense, the chair is a chair because it's not a bus, an alligator or a didgeridoo. This builds up into a collage of meaning so that a system of difference underpins the grammar, which makes possible the subtle complexities of sentences.

According to Saussure an act of speech is made possible by the structure and grammar that governs the way signs are put together. The terms he uses for this are langue and parole:

- Parole is an act of speech
- Langue is the grammar underpinning that

Parole is dependent on langue so that the meaning of individual words coheres into a structured sentence. The syntagmatic axis relates to langue and the combination/ accumulation of sign used in a sentence. The paradigmatic axis relates to the selection/substitution of different signs in parole. The sentence I climbed the tree could be extended on the syntagmatic axis to include the adjective tall: 'I climbed the tall tree'. Equally the meaning could be clarified on the paradigmatic axis, without altering the grammar (langue) by the substitution of the word tree with oak: 'I climbed the oak'. Both variations extend and refine the meaning of the sentence but in subtly different ways.

Syntagmatic

I climbed the tree
I climbed the tall tree
I slowly climbed the tall tree
I slowly climbed the tall barren tree
I slowly climbed the tall barren oak tree
I slowly climbed up the tall barren oak tree
I slowly climbed up the trunk of the tall barren oak tree

Paradigmatic

I climbed the oak
I climbed the maple
I climbed the stairs
I climbed the wall
I climbed the mountain
I climbed the pole
I climbed the ladder
I climbed the tower

What we can extract from this is a model in which the relationship between the exchange value of material goods is linked not simply to the time it took to make the item or even it's use value but its symbolic value. For example, the exchange value of a piece of wood will not simply be dependent on what sort of wood it is (maple, oak) or even what we propose to do with it (build a ladder or stairs) but the way in which we codify those things in language. Surplus value (or profit) in late capitalist society is in this sense very much connected to the relationship between the signifier and the signified. This is of course central to brand value.

The exchange value of products bearing designer labels like Louis Vuitton, Prada and Versace, for example, is based not upon 'use value' but the abstract values they signify: luxury sophistication and discernment. The interconnectedness between semiotic structure and the arrangement of capitalist power cannot be overstated and is of course the backdrop against which Valentin Volosinov writes in Marxism and The Philosophy of Language (1926).

3.3 Valentin Volosinov – Marxism and The Philosophy of Language

NAME: Valentin Volosinov (1895 to 1936)

KEY IDEA: Language is ideological: it reflects a dynamic system of beliefs or ideas; it is the key to what makes us human and the structure of the social networks we build. Volosinov views the meaning of words as arbitrary yet fluid: changing over time and according to context. He coins the term 'multi-accentuality' to descibe this. Volosinov argues that it is in the interests of the ruling class ot supress this and enforce 'uni accentuality': in effect a sign system that is more stable and fixed. For this reason Volosinov views multi-accentiality as synonymous with class resistance, which in Marxist logic leads to Communist revolution.

The past future of air travel (multiaccentuality)

The subjective nature of the signifiers we attach to whole range of objects is central to Volosinov's pre-occupation with semiotics in Marxism and The Philosophy of Language. For Volosinov language is inherently ideological; reflecting a dynamic system of beliefs; he believes the fluidity of language is central to what makes us human. Like Saussure, Volosinov views the relationship between signifier and the signifier to be arbitrary; however, unlike Saussure, Volishinov is interested in the diachronic axis: the way language changes over time as well as context. A contemporary example of this is the word 'gay', which was once used to mean 'jolly and upbeat' before becoming synonymous with homosexuality in the second half of the Twentieth Century. More recent usage of the word has shifted again: though not unconnected with cultural discrimination against non-heterosexual forms, in the early part of the Twenty-First Century 'gay' is also used as a pejorative term to describe anything deemed to be flawed, deficient or second-rate.

While this shift may be seen as culturally regressive, for Volosinov, the diachronic fluidity of the sign system is symbolic of a free society and he uses the term 'multi-accentuality' to describe this changeableness. Central to this is the notion that the meaning of a sign could be multiple according to it's context. For example the word 'there', denoted by the phonetic arrangement of 'th air' (there) can work as a noun, pronoun, adjective and adverb without even using the homophones 'their' and 'they're'. Newspaper headlines and photo captions are a fantastic example of this. For example, the narrative attached to a photograph of an aeroplane would shift greatly if it were accompanied by the caption 'doomed' instead of 'historic' or ''re-united'. The same process can be observed in non-linguistic systems. For example, upon seeing a house with a broken window and a burglar alarm going off we may well interpret that the house has been broken into. However, it could just as easily be the case that the window was broken by a tennis ball two days earlier and the owner has accidentally set the buzzer off. The key point for Volosinov, however, is that it is in the interests of the ruling class to control this multi-accentuality.

In Volosinov's logic the stabilisation of the sign system and the enforcement of uni-accentuality is key feature of capitalism. And, indeed, contemporary exchange values assigned to consumer goods are predicted on the stability of symbolic order; premium brands are in this sense very dependent upon their heritage and the stability of what they signify. Intrinsic to the symbolic value of a motoring brand like Rolls Royce for example is it pedigree: the uni-accentuality of what it means. The interconnectedness of the stability of the sign system and the cultural prerogative of the ruling class is therefore central to Volosinov's view of the multi-accentuality as a means of class resistance. And indeed, it is easy to imagine a scenario in which the hegemony of symbolic order could be disrupted. In the adverse conditions created by a natural disaster, for example, those objects that aid survival would have a much higher exchange value than items with nominally higher symbolic value. In that situation the woman with a supply of firewood and matches would be richer than a woman with a wardrobe of Gucci and Prada. By contrast, within the domain of designer clothing, an example of class resistance in Britain in the mid-Noughties was the appropriation of the Burberry by consumers from lower socio economic groupings than the label was traditionally associated. Though it is questionable how conscious this subversion of the uni-accentuality of Burberry as a symbol for the upper class was, its proliferation amongst poorer customers certainly liberated the multi-accentuality of the brands meaning.

There are of course exceptions that prove this rule. Indexical signs, for example, are signs that have some kind of direct connotation with what is being signified. Traffic signs have to operate in this way. Thus red is used to connote danger across the world, although that code also has very definite origins in nature. Likewise, iconic signs have a physical similarity to the objects they 'signify'. Their meaning goes uncontested; though arbitrary the relationship between the signifier and the signified is stable. Traditionally this term has been associated with depiction of holy figures in religious paintings and statues. In a more secular world, however, this is perhaps no longer the case and often the term is used to describe famous people. Yet, as the silk screens of Andy Warhol testify the famous are often described as icons precisely because they are so visually identifiable. In this sense there are some material values that are non-negotiable. The value of life, for example, could be said to have no price attached. Likewise, the weather, space and time are all entities, which are difficult to commoditise directly.

However, it is interesting to note that it those products and services that facilitate our access to those natural resources that often have the highest price attached to them. Holidays in the sun, labour saving devices and access to leisure space are all highly desirable products; hence a week in a seafront hotel commands a higher price than a week in a suburban travel-lodge, although the service that is being purchase may be identical or indeed less.

3.4 Vladimir Propp – Morphology of the Folk Tale

NAME: Vladimir Propp (1885 – 1970)

KEY IDEA: Unlike Volosinov and Saussure, Propp's work focuses on text level meaning and the function of narrative and character. Based on an analysis of Russsian folk-tales, Propp argues that narrative can be boiled down to 31 plot devices and commensurately only eight types of character. His work has been extremely influential upon a number of thinkers includiung Levi Straus and Roland Barthes.

KEY TEXT: Morphology of the Folktale (1928; 1968, University of Texas Press)

Another Russian theorist whose work has been particularly influential on the way in which media texts are interpreted is Vladimir Propp. Unlike Volosinov and Saussure, Propp's work focuses on text level meaning and the function of narrative and character. In line with other types of literary formalism, Propp's work is not interested in the socio-economic, cultural and historical aspects of a text, but rather emphasizes the system of grammar that underpins it. Morphology of the Folk Tale was first published in Russia in 1928, although the work was not translated into English until the 1950s. His work had an influenced on a subsequent generation of thinkers including Levi Strauss and Roland Barthes.

Propp begins his work with the assertion that folk tales are characterised by constants and variables. Variables include the 'names of the dramatis personae' while constants include their actions and functions:

> From this we can draw the inference that a tale often attributes identical actions to various personages. This makes possible the study of the tale according to the functions of its dramatis personae (Propp, 1928, 28).

If you go down to the woods today… (The Morphology of the folk tale)

Building on this idea he suggests that while the number of characters may be extremely large, the functions they perform are extremely small. Consequently, while the folk tale is superficially diverse, close inspection reveals an underlying uniformity and repetition. Key character types include the following: the villain, the donor, the helper, the princess and her father, the dispatcher, the hero and the false hero.

To understand the way in which these characterisations work in more contemporary context they are listed below relation to roles performed in Star Wars Trilogy

The villain: The Empire or 'dark side' with whom Luke Skywalker battles and Anakin Skywalker eventually resists

The donor: Obi-Wan Kenobi prepares Luke Skywalker for his odyssey and his inner journey toward Jedi-hood and gives him the gift of the lightsaber.

The helper: Yoda helps Skywalker in his quest for Jedi Enlightenment.

The princess and her father: Princess Leia is Luke's sister and Anakin Skywalker is their father. This information is concealed until later in the trilogy: revealed through Luke's use of the Jedi-force.

The dispatcher: Obi-Wan Kenobi sends Luke off on his quest.

The hero: Luke Skywalker, represents good. Does not marry the princess but she turns out to be his sister.

False hero: Hans Solo tries to marry Leia. Hans' friend Lando Calrissian is a traitor and conspires with the Empire.

Propp's work turns to focus on the analysis of narrative structure and out of which he concluded there are thirty-one plot devices used within the folk tale. These can be reduced to the following eight

1. Normality – hero introduced, member of family absents themselves.
2. Extraordinary event occurs – interdiction introduced and violated.
3. Need for vengeance
4. Magical agent is acquired by hero
5. Battle between good and evil
6. Hero escapes
7. Final task to be performed by hero
8. Glory/reward for hero.

His work has been extremely influential on the development of Media Studies and these eight stages can be applied to a range of different texts. And indeed, Star Wars conforms to Propp's blueprint for narrative. At the beginning Luke Skywalker lives happily with his aunt and uncle on the farm until The Empire kills his family while he meets with Obi Wan. Turning against the villain, Luke joins Obi Wan to avenge himself against The Empire. The gift of his fathers light sabre is exemplary of Propp's view of the magical agent. The ensuing struggle to rescue Princess Leia characterises the fight between the hero and the villain, while their reclamation of the Millennium Falcon confirms to Propp's view of the hero's escape. All that remains then is for Luke to accomplish the task of destroying the Death Star before being rewarded (ascending to the throne) at the climax of the film.

In addition to Barthes and Strauss, the influence of Propp can be felt on the French Bulgarian philosopher Tzvetan Todorov. Like Propp, Todorov work focuses on literary texts as opposed to film or television. In consideration of a range of French and American authors Todorov suggests there are two sub-genre of 'fantastic' literature: the 'fantastic uncanny' and the 'fantastic marvellous'. In the former, the fantastic occurs within the bounds of reality and a rational explanation can be found for remarkable event. By contrast the 'fantastic marvellous is categorised by a supernatural intervention. However, where Todorov's work has been most influential is in his consideration of the how genres operate. In his particular, Todorov is critical of Northrop Frye's view that in order to identify a genre an individual has to have studied every work within its corpus. Instead, Todorov argues that we should approach the study of genre scientifically. In this sense it is not necessary to have watched every Western in order to understand the constituent ingredients that go into a Western: by watching a fairly restricted selection it is inferred that there are certain unifying features common to all films within the genre e.g. the rebellious ant-hero, the desert-like setting etc. However, where Todorov is particularly influential in his assertions that there will be exceptions to the rule. Invoking the work of the scientific philosopher Karl Popper, Todorov suggested that no matter how many instances of white swans we have observed, this does not justify the conclusion that swans are white. In this sense, genre 'mavericks' like Star Wars, which reconfigures many aspect of the Western in a Science Fiction context reaffirm the central conventions of a genre.

Conclusion

In reviewing theoretical perspectives informed by semiotics and formalism, it is easy to see how they sit as cornerstones of modern thinking and can offer fascinating insights into the way in which contemporary texts make meaning. Most interesting perhaps is the tension that exists between word, sentence and text level meaning. In part, this is because it is the backdrop to contemporary ideas about the role of the audience and the function of genre. In particular Volosinov's perspective on how the meaning of texts change over time is particularly instructive in a digital age in which the archives of media production are more accessible that ever before. The effect of this on contemporary ideas about genre and narrative is complex. On the one hand, the stability of certain conventions has been reinstated because we are more familiar with older texts. On the other hand, the proliferation of narrowcast digital media means that producers have the opportunity to be more experimental in their development of new media forms.

4. The Frankfurt School and Neo Marxism

Introduction

In this chapter we look at the way in which three neo Marxist theorists can be to used to frame and shape the way in which we think about contemporary media, society and culture. The chapter begins with an overview of the Theodor Adorno's work and the way in which this can inform our understanding of contemporary media texts. The second section turns to focus on Herbert Marcuse. Finally, in the concluding part of the chapter we turn to the work of the Italian theorist Antonio Gramsci, whose writing focuses on the concept of the hegemony, and consider the ways in which this has informed cultural theory. First, however, we turn to what is actually meant by the term Frankfurt School.

4.1 The Frankfurt School

The Institute of Social Research at the University of Frankfurt am Main, or Frankfurt School as it became known, was founded in 1923 by Felix Weil. Weil was a young intellectual and one of a number of other Marxist thinkers who recognised that Marxism was impractical in the form proffered by the Communist Party and therefore needed considerable revision before it could be implemented successfully. Weil realised that in order to do this he would need to establish an academic institution independent of Communist Party funding. Weil used money from his father's business to set up the school and campus, which was affiliated with the University of Frankfurt. Alongside this Weil also applied to the Ministry of Education for a professorship in order for the school to gain university status.

The Institute was highly successful, with a number of academics making significant contributions towards moving Marxism beyond 'dialectical materialism'. However, the school was dissolved when Hitler came to power in 1933, with many of the group relocating to New York where the Institute was reformed as an affiliate to Columbia University (N.Y.C.). Theodor Adorno, perhaps the Institute's most famous social theorist, returned to Germany with his colleague Max Horkheimer in 1949 after spending four years at the University of Oxford and a further ten years teaching in the United States. Adorno and Horkheimer reformed the Institute of Social Research in Frankfurt, ushering in a new wave of post-Hegelian thinkers.

The Frankfurt School saw popular culture as an agent of class domination and capitalism: pacifying the Proletariat into accepting the inequalities of the class system. Frankfurt School theory is often referred to as critical theory because it is questioning of mass media, modern culture and existing social theory. The work of the Frankfurt School has been particularly influential in shaping the way in which mass culture has been theorised. In particular the work of Adorno and Marcuse views the media as manipulating the Proletariat, distracting workers from their disadvantaged social position and pacifying them with products that meet needs that have been falsified. In short the concept of mass media is used to explain the reasons why the revolution Marx predicted did not materialise.

The production line (The Culture Industry as Mass Deception)

4.2 Theodor Adorno and Max Horkheimer – The Culture Industry

NAME: Theodor Adorno (1903 – 1969) and Max Horkheimer (1895 –1973)

KEY IDEA: Adorno and Horkheimer criticise the culture industry, which they view as having standardised the mores of civilisation. Central to this deception is the notion that the culture industry serves the interests of the consumer; this they argue is incorrect. For Adorno and Horkheimer, the culture industry serves only the interests of the capitalism and by extension those of the ruling class. In particular, Adorno and Horkheimer bemoan the pseudo-individualisation of mass culture (films, music and popular fiction) and the distinction between the different forms of mass culture, which they argue is illusory: concealing its underlying formulaic quality. However, perhaps Adorno and Horkheimer's key point is on the function of the culture industry, which they perceive to be the distraction of the workers from the inequalities of the class system under-capitalism and their potential to revolt.

KEY TEXT: 'The Culture Industry as Mass Deception' (1944).

Adorno and Horkheimer begin 'The Culture Industry as Mass Deception' with a rejection of the sociological perspective of mass-culture as chaotic. Instead, they argue it is uniform and standardised: 'under capitalism all mass culture is identical' (Adorno and Horkheimer, 1944, 1037). Mass culture, according to Horkheimer and Adorno is characterised by few production centres and many points for consumption. In essence, what they are talking about are the mass-production methods that have characterised mechanised production since the Industrial Revolution at the end of the Nineteenth Century and typified by the methods of Henry Ford and assembly line production. However, where they differ from previous generations of neo-Marxists is their emphasis on cultural as opposed to commodity production. Adorno and Horkheimer view the culture industry as deceitful on a number of levels. In the first instance they suggests that the culture industry claims to satisfy the needs of the consumer; this they argue is incorrect. They give the example of the way in which classical music and great literature is re-appropriated in the popular domain in films and jazz, arguing that this is not in the interest of the audience but simply the need to culture industry to make profit.

For Adorno and Horkheimer, the culture industry serves only the interests of the capitalism and by extension those of the ruling class:

> No mention is made of the fact that the basis on which technology acquires power over society is the power of those whose economic hold over society is greatest. (Adorno and Horkheimer, 1944, 1037)

In this sense, Adorno and Horkheimer are re-stating one of the central tenets of Marxist ideology: that the economic is the rule and base of society. For Adorno and Horkheimer, movies occupy the same role as automobiles and bombs: they 'keep the whole thing together'; concealing from the Proletariat their disadvantaged class position. Indeed, they contend that the separateness of the culture industry from other forms of capitalist monopoly is illusory:

> [S]teel, petroleum, electricity, and chemicals. Culture monopolies are weak and dependent in comparison… The dependence of the most powerful broadcasting company on the electrical industry, or of the motion picture industry on banks, is characteristic of the whole sphere. (Adorno and Horkheimer, 1944, 1038).

43

Symptomatic of this is not only the way in which the products of the culture industry are gradated according to models of economic efficiency, but also the manner in which audiences are demarcated and isolated. However, it is perhaps the illusion of difference within the culture industry itself that is perhaps the most persuasive argument put forward by Adorno and Horkheimer.

Central to Adorno and Horkheimer's thesis is the notion that the culture industry offers products that are misleading in their difference. Just as the difference between a car produced by Chrysler and one produced by General Motors is fairly minimal so too is the notion of choice between the films of Warner Brothers and Metro Goldwyn Mayer deceptive. Indeed, Adorno and Horkheimer contend that 'connoisseurs' who discuss relative worth of culture industry products are actually perpetuating its myth. In particular, they bemoan the use of stylisation to create the illusion of difference. Though they concede that style has a place in the authentic art of the Enlightenment period, in the products of the culture industry Adorno and Horkheimer view style as a weak substitute for true aesthetic value. Unlike art that transcends reality, stylised mass media is viewed as a conduit for ensuring society's obedience to existing social hierarchy.

> By occupying men's senses from the time they leave the factory in the evening to the time they clock in again the next morning with matter that bears the impress of the labour process they themselves have to sustain throughout the day, this subsumption mockingly satisfies the concept of a unified culture which the philosophers of personality contrasted with mass culture. (Adorno and Horkheimer, 1944, 1041)

In this sense, 'The Culture Industry as Mass Deception' conforms to a classic Frankfurt School perspective in that it views the function of the culture industry to be the distraction of the workers from the inequalities of the class system under-capitalism and their potential to revolt.

One of the areas of which Adorno was most critical is popular music. In 'On Popular Music' (1941) he argues that pop music is formulaic and standardised. And, certainly when one considers the popular music of Thirties and Forties, it is easy to see why. As Howard Goodall describes in Twentieth Century Greats (Channel 4, 2004), by the mid-Twentieth Century there was an impasse between classical 'art' music and popular forms. While the avant-garde of the period challenged Western forms of musical composition, popular songs were unadventurous and prescribed. Likewise, the domination of the contemporary in the Noughties by reality based talent shows like Pop Idol and X-factor reinforce the idea of popular music as pseudo-individual and mechanistic. Indeed, the notion that programs that embody the Warholian mantra that everybody can be famous for fifteen minutes distract individuals from the decline of class mobility and the polarisation of wealth and opportunity in the Noughties, is very real possibility. In this sense the critique of Frankfurt School theorists remains very relevant, particularly in light of the capitalist business infrastructure underpinning the entertainment industry.

The flip side of this viewpoint is that Frankfurt School critiques fail to grasp the complexity of the way in which commoditised cultural forms make meaning. And certainly, from a sequential point of view, Adorno and Horkheimer's critique of popular music predates the transformation of popular music culture in the 1960s. In this sense, it could be argued that the way in which popular music embraced aspects of avant-garde musical in the wake of the Beatles and the British invasion of the American charts in the 1960s demands a re-think of this perspective. Moreover, the way in which popular music culture embraced an art-school aesthetic, particularly in its rehabilitation of the Pop Art sensibility, challenges the views of consumer culture as facile and one-dimensional. Typical of this is the legacy of Richard Hamilton and the Independent Group. Though their exhibition This is Tomorrow (1956) precedes the cultural revolution of the 1960s, the group's influence can be felt on a generation of musicians. Definitive of this trend is the legacy of 1970s rock group Roxy Music. Unlike their forebears, Roxy Music not only used the studio as an instrument, but also embraced their own commoditisation as medium of aesthetic exploration. In this sense, the glamour of surface culture and the allure of capitalist modes of dissemination are viewed as part of the conceptual frameworks within which popular music makes meaning.

4.3 Herbert Marcuse: One Dimensional Man

NAME: Herbert Marcuse (1898 – 1979)

KEY IDEA: Marcuse argues that man is not able to imagine a way of living other than life under capitalism. For this reason he argues that man is one dimension. In particular, he views capitalism as oppressive of the Art, the duty of which he believes is to expose the contradictions of the capitalist system. Commoditised cultural forms are in this sense extremely problematic. Unlike Marx, Marcuse is therefore extremely pessimistic in his view of the working class and their potential for revolutionary action.

KEY TEXT: *One Dimensional Man* (1964).

False needs never met (One Dimensional Man)

Like Adorno and Horkheimer, Herbert Marcuse's work is extremely critical of modern industrial society. Though Marx influences him, he views the capitalist system as so strong that it is impervious to revolutionary action. Central to the strength of the capitalism, according to Marcuse is that man is no longer able to imagine another way of living; it is in this respect that he argues man has become one dimensional. An area of particular concern for Marcuse is creativity and the arts. In Twentieth Century capitalist society these have become incorporated in to the system of the ruling class. In particular he finds the commercialisation of aesthetic forms to promote capitalism extremely problematic. For Marcuse, the arts should be sceptical if not critical of capitalism, not co-opted into the endorsement of commoditised cultural forms, profit and exploitation. In particular he views those artists whose work appears to criticise capitalists systems as particularly problematic: more often than not these artists are contractually involved with corporations whose purpose is to make profit. In this sense, the commoditisation of the arts under capitalism is in effect a deterrence mechanism; for all its revolutionary potential, in most instances it exists to distract society from the inequalities of the class system.

One area in which Marcuse is substantially different to Marx is in his view of the working class. While Marx views them as potential agents of communist revolution, writing over one hundred years later, Marcuse is less optimistic. He contends that the working class have been corrupted by capitalism and the allure of mass culture. Central to this is the fact that living standards have risen in the Western World under capitalism. In this sense Marcuse argues that modern society is determined by 'false needs':

> The people recognise themselves in their commodities; they find their soul in their automobile, hi-fi set, split level home, kitchen equipment. The very mechanism which ties the individual to society has changed; and social control is anchored in the new needs which it has produced. (Marcuse, 1964, 9).

Marcuse's influence can be felt most obviously in the work of theorists who explore the role of consumer culture in capitalist society. In particular, Jameson and Baudrillard's take on commoditised cultural forms is rooted in the belief that they appeal to false needs. Where they modify Marcuse's perspective is in the belief that the purchase of material goods is connected to the signification of identity and in this case an active rather than passive process. By contrast, for Marcuse, mass culture pacifies:

> (T)he irresistible output of the entertainment and information industry…promote a false consciousness which is immune against its falsehood… it militates against qualitative change. (Marcuse, 1964, 26-27)

However, perhaps where Marcuse's work is still relevant is in his explanation of why society has not embraced the revolution Marx predicted. Capitalist modes of production in the West have pacified the Proletariat with rising standards of living and cheap consumer goods. However, it is perhaps the contradiction he perceived in the commoditisation of aesthetic cultural forms that is most instructive.

Marcuse emphasizes the endemic ideological contradiction in the commoditisation of cultural forms that are explicit in their critique of capitalism. On the one hand, the status of art as product negates the power of that censure. For example, how can a novel or play published for profit be condemning of capitalist systems? On the other hand, if it is the responsibility of the arts to be critical of the exploitative nature of capitalism the context of exchange and distribution need not be made problematic. Exemplary of this dichotomy was the chart battle for Christmas number one in the UK at the end of 2009. On the surface, the choice between X-Factor's Joe McElderry 'The Climb' and the Facebook generated campaign for the reissue of Rage Against the Machine's 'Killing in the Name of' seemed to represent a clear choice between commoditised mass culture and a less contrived and more authentic aesthetic expression. However, as the popular press noted the time, the choice was fairly illusory: both acts were signed to the Sony Music Corporation of America.

4.4 Antonio Gramsci – Hegemony, Intellectual and the State

NAME: Antonio Gramsci (1891 – 1937)

KEY IDEA: Central to Gramsci's view of social structure is the notion of 'hegemony': the supremacy of one social grouping over another. Hegemony asserts itself in two ways: domination and moral leadership. For dominance to be effective, however, it needs to lead first. Consent is, in this sense a key concept. Agents of cultural consent include media industries like the popular press, film, television and pop music.

KEY TEXT: 'Hegemony, Intellectuals and the State' (1971).

Like Adorno, Horkheimer and Marcuse, Antonio Gramsci was heavily influenced by the work of Marx and Engels. Indeed, Gramsci was, for a time, the leader of the Italian Communist Party and was imprisoned under Mussolini's fascist regime in 1926. In particular Gramsci views the role of the economy as pivotal to social changes. However, unlike Marx and Engels, Gramsci contends that ideas and beliefs could change the economic structure of society. In this direction, Gramsci was interested in how the working class could develop their ideologies of resistance to challenge the inequalities of the capitalist system.

Central to Gramsci's view of social structure is the notion of 'hegemony'. Put simply 'hegemony' refers to the supremacy of one social grouping over another. This, Gramsci contends, asserts itself in two ways: domination and moral leadership. For dominance to be effective, however, it needs to lead first. Consent is, in this sense a key concept. In 'Hegemony, Intellectuals and the State', written while Gramsci was in prison and first published in 1971 he states:

> The 'normal' exercise of hegemony on the now classical terrain of the parliamentary regime is characterised by the combination of force and consent, which balance each other reciprocally, without force predomination excessively over consent. Indeed, the attempt is always made to ensure that force will appear to be based on the consent of the majority, expressed by the so-called organs of public opinion – newspapers and associations – which, therefore, in certain situations, are artificially multiplied. (Gramsci, 1971, 80)

While Gramsci concedes that the hegemony has to take on board the interests of groups over which hegemony is exercised it must also always serve the 'decisive nucleus of economic activity' (Gramsci, 1971, 161). Gramsci uses the term hegemony to describe not only the relationship between different economic groupings within society, but also international relationships.

The notion of hegemony remains a highly influential concept that can be dislocated from spheres of culture explicitly connected with materialism and capitalist acts of material exchange. Many feminists use the term to describe patriarchal structures of society and the subordinate position of women in Western culture. Likewise, the term has been appropriated by post-colonial theorists, who view the power dynamic between the West and the developing world in terms of cultural imperialism. Within a British context the most explicit example of the way in which the mass media is an agent of cultural consent is the influence of the popular press on voting behaviour. Particularly influential is The Sun, Britain's most widely read paper, owned by Rupert Murdoch's News International. In the past twenty years the title has switched allegiance from the Conservative Party in 1992, to Labour in 1997 and back again to the Conservatives in 2009. At each successive turn this support has pre-empted a shift in the electorate from the right to the left and back again. It will, therefore, be fascinating to see what the outcome will be of the prospective 2010 election.

Another area in which Gramsci's work that is particularly influential is on the role of the intellectual in capitalist society. Gramsci suggests that the specialist intellectual is a feature of capitalism, serving the systems need to understand its own function economically, socially and politically; hence the pre-eminence of sociologist Anthony Giddens' role as advisor to the former Prime Minister Tony Blair. However, Gramsci's view of intellectualism is holistic: 'all men are intellectuals... but not all men have in society the function of intellectuals' (Gramsci, 1971, 9). Indeed, Gramsci, contends, there is no such thing as a non-intellectual. However, the specialisation of the intellectual is a function of hegemony because conquering the ideology of preceding intellectuals is a key feature of any group working towards dominance and hegemony. In this direction Gramsci contends that the intellectual has two key roles: the acquiescence of spontaneous consent and the enforcement of legally coercive power. It does not automatically follow, however, that successive generations of intellectuals will be wholly successful in conquering the preceding cohort. Gramsci gives the example of the intellectual class emanating from Britain's land-owning aristocracy, which maintains an intellectual monopoly in spite of its loss of economic supremacy.

The Acquiescence of Consent (Hegemony)

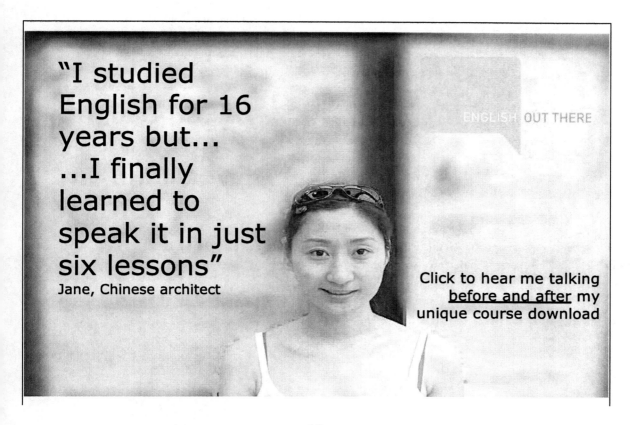

The politics of education is a typical example of this tension. On the one hand, for many, education should adhere to epistemological demands of discrete subject specialisms. Laid over this is the informal or 'hidden curriculum', which reinforces many of the primary functions of socialisation carried out in the family. On the other hand, education has to adhere to the demands of economy and indeed the proliferation of state education at the end of the Nineteenth Century was in this sense a function of the need for a more literate work force. Likewise, successive education policies, which have seen the school leaving age raised from 14 to 16 to an incumbent 18 reflects the need of the British economy for a more educated work force. The tension between these two schools of thought is played out most memorable in Alan Bennett's The History Boys (2006) in which a pair of maverick schoolmasters embody two very different ideologies of education. Hector symbolises a belief in education for its own sake, while Irwin views the exam system more strategically: a game that his students must learn how to play (and win).

An aspect of British education that is particularly interesting in this respect is the British public school system. As Britain embraced comprehensivisation in the 1960s and 70s, few would have believed that private schools would have seen a resurgence in the 1980s. While this was a reflection of Conservative Party policy that encouraged parents to sponsor assisted places, the success of public schools in the Twenty-First Century is a legacy of the intellectual monopoly of the Britain ruling class from the days of Empire. The prestige of British education on the world stage is such that while some places at public schools still go to the children of landed gentry and well-healed Brits, an increasingly large proportion of students at schools like Harrow, Radley and Eton are the children of newly wealthy over-seas parents.

The third area in which Gramsci is particularly influential is the way in which he conceptualises the role of the State. In the first instance Gramsci argues that the State is ethnical because one of its most important functions of raising the cultural and moral standing of the population to a level that corresponds with the interests of the ruling class. Education is in this sense a key function of the State, not just in terms of the skills required of workers but also in terms of culturally consent to hegemony. Central to this is the fluidity of membership of the bourgeoisie or ruling hegemony:

> The previous ruling classes were essentially conservative in the sense that they did not tend to construct an organic passage from the other classes into their own i.e. to enlarge their class sphere 'technically' and ideologically: their conception was that of a closed caste. The bourgeois class poses itself as an organisation in continuous movement, capable of absorbing the entire society, assimilating it to its own cultural and economic level. The entire function of the State has been transformed; the State has become an 'educator' (Gramsci, 1971, 260).

The State, in Gramsci's logic, however, is more complex than it is for Marx. Rather than viewing the state as a just cipher for capitalist exploitation, Gramsci views the state as a multifaceted institution that not only maintains dominance but also manufactures acquiescence and consent.

The transformation of the public sector in the Britain during the 1980s is exemplary of this trend. Part of Margaret Thatcher's economic policy was the pursuit of fiscal expedience on public spending: funding was cut across a range of public sector services including the social housing, public transport and defence. In addition to this one of the central policies of what became known as Thatcherism was privatisation: the selling off of state utilities to private companies. To illicit consent for social policy, which involved selling off public owned assets, Thatcher and her government reconfigured the electorate as consumers of state services. In particular, the changes to state education in the wake of the 1988 Education Act encouraged parents to make comparative consumer choices about their children's education based on league tables and independent reports. Likewise the sale of council housing stock to existing tenants in social housing embourgeoised a demographic of working class voters who now stood to benefit from lower taxes and a decrease in public spending.

For Corner and Harvey, the notion of heritage was central to the way in which the government procured the acquiescence of cultural consent. Writing in 1991 they state:

> History is gradually being turned into something called heritage whose commodity value run from tea towel to the country house. Its focus on an idealised past is entropic, its social values are those of an earlier age of privilege and exploitation and it serves to preserve and bring forward into the present. Heritage is gradually effacing history, by substituting an image of the past for its reality. (Corner and Harvey, 1991,

In the arts this shift towards privatised conceptions of nationhood is manifest in heritage films like A Room With a View (1985) and Howard's End (1992): literary adaptations typified by Merchant Ivory productions. Such films are, according to Andrew Higson, synonymous with specific narratives of historical representation:

> The heritage films… work as pastiches, each period of the national past reduced through a process of reiteration to an effortlessly reproducible, and attractively consumable, connotative style. (Higson, 1993, 112)

A manifestation of this in the 'heritage film' he suggests is a pictorial camera style, which emphasizes the glorious and fetishised surroundings: 'the creation of a heritage space, rather than a narrative space' (Higson, 1993, 117). Commensurate with this then is the tacit invitation to consume.

A more contemporary example of the way in which media texts legitimise hegemony is that of the reality television. Programs like Big Brother and talent shows like the X-Factor, which manufacture 'celebrities', taking ordinary people and making them famous, help maintain an illusion that Britain is a socially mobile country. In real terms social mobility has declined in the last 20 years with many not achieving the same level of professional and education success as their parent. And, indeed, when one considers the level of voter apathy in the Noughties it would seem that consent to the ruling hegemony has never been greater. For example, in the 2001 and 2005 elections 41% and 39% of people did not bother to vote (www.ukpolitical.info): a figure that has risen by nearly 18% since the 1990s (electoral turn out was 77% and 71% in the 1992 and 1997 elections). What this says about British politics is of course open to interpretation; however, it is interesting to note that this period of apathy has coincided with the proliferation of the Internet and the revolution in digital broadcast media. By contrast, the years in which electoral turnout was highest (84% in 1950 and 83% in 1951) preceded the proliferation of the television in Britain in the wake of the Coronation in 1952.

This is of course a crude barometer and there are multiple factors that effect voting behaviour. Moreover, the increased levels of interactivity offered by new media offer much greater opportunity for democracy than traditional means of political expression. However, with each successive development in media technology, political agencies are becoming ever more sophisticated in their use of the media. This can be as simple as the way in which those in charge of George W Bush's wardrobe in the run up to the 2000 General Election ensured that their candidate was seen dressed in clothing that would appeal to the swing voter. Or it could be the way in which Barack Obama's 2008 campaign for the White House emphasised the grass-root support evidenced by small-scale Internet donations. Both strategies emphasise consent as opposed to submission. It is perhaps for this reason, therefore, that of all the strands of neo-Marxism discussed in this chapter, Gramsci's view of hegemony has most to offer contemporary media analysis.

Conclusion

In reviewing the work of three neo-Marxists it is easy to see how they have informed modern thinking and can offer fascinating insights into the way in which the media industry operates. Most interesting perhaps, is the tension that exists between the pleasure of consuming mass culture and the ideological compromise this entails. In part this is because this debate is the backdrop to current thinking about the exploitative nature of global capitalism. Though critics of neo-Marxist thinking would argue that it misunderstands the creative and reflexive ways in which the individual engages in mass culture, the concept of hegemony remains very persuasive. In particular the notion that commoditised cultural forms are less than innocent has remained a recurrent theme in critiques of mass culture. Moreover, the notion that the passive consumption of media and other cultural products acquiesces consent to capitalist power requires little clarification.

5. Structuralism

Introduction

This chapter focuses on structuralism and how it has influenced the ways in which we view the forms and conventions of cultural texts. In particular we explore how human culture is understood as a system of signs reflecting 'deep structures' that shape the 'common sense' ways in which we interpret texts. We begin with an exploration of the work of Levi Strauss, focusing in particular on his use of binaries. We then turn to look at how Jacques Lacan used this systematic approach to Psychoanalysis. Finally, in the concluding section we focus on the work of Stuart Hall and the way in which this has framed and shaped the development of Cultural Studies. First, however, we will consider what is meant by the term structuralism.

Structuralism

Structuralism is closely connected to semiotics in that it views society and culture as a system or code. In this sense it draws heavily upon the work of Ferdinand de Saussure, outlined in Course in General Linguistics (1916) and discussed in chapter 3. Likewise there are elements of Russian formalism to structuralism. In particular the work of Vladimir Propp, discussed in chapter 3 was a great influence on Levi Strauss. However, where Strauss's work differed from other theorists influenced by Saussure and Propp is in his application of structured theory to cultural phenomena, as opposed to pure linguistics and literary texts. Strauss was one of a generation of sociologists who looked at the way in which binaries operated in society. In particular he viewed the social world to be constructed of opposites: hot and cold, raw and cooked, male and female. Others like Lacan appropriated these structures into Psychology, arguing that the human mind operates in ways that are systematic and regulated. Though structuralism is often seen as being too deterministic in its approach to cultural phenomenon, it had a considerable influence on the work of Barthes and Derrida and indeed, perhaps the greatest legacy of structuralism is the begetting of its antithesis in the form of post-structuralism. Indeed, for many postmodern theorists and feminist critics structuralisms reliance upon Enlightenment ideas about scientific knowledge and the application of logic render its contemporary usefulness limited. That said, it is arguable that the area in which structuralism has had most influence has been in Cultural Studies and, in particular, the work of Stuart Hall.

5.1 Levi Strauss – Cultural Semiotics

NAME: Claude Levi Strauss (1908 – 2009)

KEY IDEA: Building on the foundational work of Ferdinand de Saussure, Strauss developed the linguistic science of semiotics to encompass broader anthropological concerns of understanding the cultures and myths of tribal societies as means of uncovering the universal or 'deep' structures of society. In particular, he emphasises the relevance of cultural binaries in revealing over-arching systems and structures.

KEY TEXT: *The Savage Mind* (1963).

Claude Levi Strauss was a French anthropologist born in 1908. His work has provided the blueprint for textual analysis of cultural and media texts that remains relevant - in spite of the advent of postmodernism - to this day. In part this can be attributed the systematic way in which it enables the reader of the text to extrapolate meaning from the seemingly discordant and chaotic. However, Straussian analysis also paved the way for poststructuralism, which far from being an outright rejection of structuralism is in fact a revision or extension of it.

Building on the foundational work of Saussure, Strauss developed the linguistic science of semiotics to encompass broader anthropological concerns. In particular he focused on understanding the cultures and myths of tribal societies and viewed this as a means of uncovering the universal or 'deep' structures of humanity. Indeed this essentialist approach of attempting to homogenise meaning was the main criticism of Strauss' work by poststructuralists like Jacques Derrida and Michel Foucault. A further criticism stems from Strauss' reliance on binary oppositions to define cultural differences. One such example is Strauss' 'culinary triangle' in which he describes graphically, the relationship between culinary tradition and distinction between nature and culture as governing forces within societies. Strauss suggests that the way in which raw food is transformed for preservation relates directly to whether or not a particular society is founded on a belief in nature or culture. In his logic when rations are left to ferment, rot or dry this is symptomatic of society that believes in nature. Conversely, when raw food is transformed by cooking this implies that the society is founded on a belief in culture.

Most of Strauss' anthropological work focuses on the stories and myths of what were then referred to as primitive societies. In this direction he argues that whilst myths contain many contingent elements that could be seen as culturally or geographically specific, there are also common features. These, according to Strauss, can be isolated when a number of myths are 'de-cluttered' of that which is purely arbitrary and instead compared 'laterally'. By focusing on the commonalities of myths, predominantly binary oppositions, Strauss believed that he could expose the universal structures of the human psyche. For Strauss 'mythical thought always progresses from the awareness of oppositions toward their resolution' (Strauss, 1963, 224). Where Strauss's work is particularly useful, however, is the way in which he extends this analysis towards everyday social relations, focusing specifically on kinship relationships across many different cultures. For example, in The Savage Mind (1963) he focuses on the 'grammar' of interpersonal relationships in his quest to isolate the deep structures social bonding. In this study Strauss identifies four main permutations of interaction within the family:

Marriage (husband-wife)

Siblings (brother-sister)

Filiations (parent-child)

Avuncular (uncle/aunt-nephew/niece)

Parallel Lines (Cultural Semiotics)

He argues that by identifying the totality of isolated/combined permutations we are able to predict how a slight difference in one relationship will affect the 'form' the others will take. For example, if child has only one parent (i.e. normative filial relations are disrupted) then the role of an uncle or aunt (the avuncular) might become more important. By the same token, if an uncle and aunt (the avuncular) is unable to have children the role of the niece or nephew may become more imperative thereby reinforcing a relationship with a sibling.

Structuralist textual analysis of the cover of Parallel Lines by Blondie

The stark contrast between black and white is the main focus of opposition within the image. This exists not only in the striped backdrop but also in the dress codes of the band, most notably in the contrast between the black ties and white shirts of the male figures and in the blonde peroxided hair and dark roots of the female. These colours also connote gender roles; the men are dressed in black formal attire symbolising power and status, while the woman is a wearing white dress which, whilst instinctively symbolising the ownership and the objectification by the masculine gaze, simultaneously unfurls further binaries: the formal (suits) and the casual (the resemblance of the dress to a 'baby-doll nightie') in turn revealing the virginal and the sexual. In contrast to this, is the visual hierarchy of the image, which privileges the groups singer Deborah Harry by placing her centrally and in front of the men; symbolising strength and control while the men seem a little emasculated and boyish. The clothing worn also reflects this idea. The scruffy soiled trainers that the men wear with their suits together with their laid-back posture, conflicts with the image of masculine status, creating the impression of juvenile irresponsibility that contrasts with the matronly impression that the below-the-knee-hemline of Harry's dress creates. Likewise on the one hand Harry's hair colour and style connotes '50s Hollywood film stars like Marilyn Monroe, thus assimilating the codes of masculine objectification; whilst on the other hand these masculine consumer ideals are subverted by the singer's dark roots which make her hair look 'trashy' and the white bandage on her arm which denotes her as a drug user.

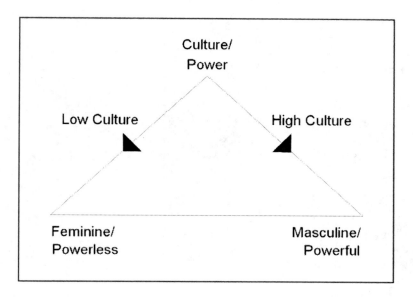

Straussian triangular analysis of Parallel Lines cover

Strauss' method of semiotic analysis has been very durable and is still used as the primary means of analysing texts within Cultural and Media Studies disciplines. In Strauss' logic the structures of culture are not readily observable because we accept them in an unquestioning way as 'common sense'; therefore we must use structural analysis in order to uncover them. For Strauss, this instinctive categorisation of cultural phenomenon enables us to use binary oppositions. Focusing on the cover to Blondie's Parallel Lines (1978) a number of categories reveal themselves: black/white, male/female. The significance of this for Strauss, however is that these binaries reveal structural categories like colour and gender. However, identifying further binaries refines analysis within these parameters. Having identified the structural category of gender, for example, we 'naturally' isolate further binary oppositions between femininity and power, revealing more subtle binaries within gender roles. It is to the construction of gender that we now turn in our consideration of the way in which Lacan applies structuralist analysis to psychoanalysis.

5.2 Jacques Lacan – The Mirror Stage

NAME: Jacques Lacan (1901 – 1981)

KEY IDEA: Through a process of narcissistic self-objectification the ego is formed. This encounter is determined by the same libidinal drives as Freud's 'Phallic Stage' of psychosexual development. However, it produces a 'misunderstanding' in which the imaginary is mistaken for the real, thus producing the ego. The ego is produced via language in the symbolic order; the mirror image becomes not only a symbol for the unified subject but also a signifier for the self; a signifier for you-ness.

KEY TEXT: 'The Mirror Stage' (1937).

Who am I? (The Mirror Stage)

The French psychoanalyst Jacques Lacan was born in Paris in 1901. He trained in medicine and psychiatry and in 1931 began working as a forensic psychiatrist whilst completing his doctorial thesis. When this was published in 1932 it found little acclaim in psychoanalytical circles; however, it did attract interest from members of the surrealist Art movement. In 1934 Lacan became a candidate for the Paris Psychoanalytical Society. It was around this time that he met and befriended the surrealists André Breton and Georges Bataille. Lacan developed a close relationship with the surrealists having both an influence upon the movement and in turn, being influenced by it. Salvador Dali's painting The Metamorphosis of Narcissus (1937), for example, explores the mismatch between the corporeal human body and its idealised image mediated via sexuality and the ego; a concept owing much to Lacan's report on his first major psychoanalytical study into what he terms the 'Stade du Miroir' or 'Mirror Stage' of child development. Drawing upon Sigmund Freud's work on the Oedipal nature of psychological development in male children, Lacan explores the possibility of a rift or mismatch between physiological and psychological states at an early stage of child development; a disparity, which characterises the division between subject and object in the Eighteenth Century Enlightenment model of selfhood. Lacan refers to this stage of development as the 'Mirror Stage'.

The 'Mirror Stage' refers to the moment when a child first recognises itself in a mirror. At this point the child has not yet attained full coordination or control of its body. Until this point, the child experiences its corporeal (bodily) self only through proprioception, or the perception of the spatial and sensory relationship between neighbouring body parts and the physical environment. The child's sense of self is discordant, un-unified and fragmented, and yet, in the mirror he or she identifies their own body image as a unified whole. Initially the child tries to verify the image as its own by turning to the mother, whom in Freudian terms is not only biologically unified with the child, but wedded to its id, or primal self as expressed in its physiological drives. In doing this however, the child recognises itself as a being separate from its mother. And, as if to test this newfound autonomy, the child then begins to explore its movements as they are reflected in the mirror; thus identifying with the idealised and unified mirror-self. For Lacan, it is through this process of narcissistic self-objectification that the ego is formed. He further suggests that this encounter is determined by the same libidinal drives as Freud's 'Phallic Stage' of psychosexual development. Furthermore he argues that the encounter produces a 'misunderstanding' in which the imaginary is mistaken for the real, thus producing the ego. The ego is produced via language in the symbolic order; the mirror image becomes not only a symbol for the unified subject but also a signifier for the self; a signifier for 'you-ness'. The mirror self is not a decentred and fragmented postmodern self, but the idealised self of the Enlightenment.

An interesting example of how the mirror stage functions within contemporary media is the social networking website Facebook. Facebook exemplifies the way that new media can be used as a 'mirror' in front of which we are able to reflexively re-present and reproduce the self. The Facebook profile is a phenomenon, which has arguably fundamentally altered the nature of social interaction. Not only is it possible to "keep in touch" with numerous familiar others without the commitment of a phone conversation, letter or even an email but also it is possible to represent one's own life as what is effectively a media product. Facebook 'friends' are arguably subscribing to a niche brand – the brand of you! This logic has arguable redefined the function of social interaction from a concerned wish to find out about the wellbeing of others, to a narcissistic want for self-affirmation of one's own 'brand identity' via the 'feedback' of others. In effect Facebook is operating in the realm of regulatory systems or what is commonly known as the theory of cybernetics. Facebook and many other Internet systems operate on the basis that the communications loop between output (from the sender) and feedback (from the receiver) is regulatory. In other words the feedback received in return for output has the effect of reshaping future output. Therefore the ways in which a Facebook user projects themselves into cyberspace is a contrivance or reformation of self, orientated towards his/her audience and designed to solicit affirmative feedback. This can be likened to Lacan's observation that upon initial (mis)recognition in the mirror, a young child will look to its mother for affirmation of its own existence.

5.3 Stuart Hall – Encoding and Decoding

NAME: Stuart Hall (1932 to present)

KEY IDEA: Proponent of audience reception theory, Hall looks at the way in which cultural interaction generates consent for hegemony (a term he borrows from Italian theorist Antonio Gramsci) – the dominant ideology of the ruling class. Hall views audiences as both the producers and consumers of texts: decoding the meaning encoded by the originator of the text. His approach to textual analysis is that the consumer actively negotiates the meaning.

KEY TEXT: 'Encoding/Decoding' in *Culture, Media, Language* (1973).

Stuart Hall's work follows in the footsteps Antonio Gramsci and Louis Althusser. And, like Althusser, Hall could be described as a Structural Marxist in the sense that his work sets out to uncover the ideological structures within media representations of reality. Janet Woollacott argues that Hall's analysis of the media's signification practices adopts Althusser's notion of the media 'as an ideological state apparatus largely concerned with the reproduction of dominant ideologies' (Woollacott 1982: 110). However, Hall also challenges the traditional concept of mass communication as the linear route between sender/message/receiver, and in doing so opens up the possibility for consumer resistance. In other words his theory suggests that consumers of media texts produce meanings other than those that are intended or 'preferred' by the media producer.

Encoding/Decoding (1973) explores the structural mismatch between sender and receiver. Hall argues for a new model of the mass communications process as a structure produced and sustained through the articulation of linked but distinctive moments – production, circulation, distribution/consumption, reproduction (Hall, 1973, 128- 138). He places particular emphasis on the audience and the relationship between structure and agency in which he challenges previous conceptions of audience as a 'passive' element within an uninterrupted circuit of communication'. Shannon and Weaver (1949), for example, produce a model, which doesn't allow for the influence of social contexts in 're-producing' meanings in decoding the message:

1. An information source, which produces a message.
2. A transmitter, which encodes the message into signals
3. A channel, to which signals are adapted for transmission
4. A receiver, which 'decodes' (reconstructs) the message from the signal.
5. A destination, where the message arrives.

NOISE interference with the message travelling along the channel, which may lead to the signal received being different from that sent.

From Shannon and Weaver (1949), Transmission Theory of Communication.

Noise is described as that which interferes with the message as it travels along its channel: for example, static interference on a telephone or TV signal. Thus it is seen as external to an otherwise unhindered process. Hall however, argues that whilst this model is fine for interpreting communications technology, it does not take into account the fact that each stage of the process exists as a discreet moment within a social context. Furthermore, he suggests that both the transmitter/encoder and receiver/decoder are shaped the most by this discursive dimension, and are therefore the most important elements within the chain. He argues that events cannot be 'transmitted' in their existing form, only 'signified within the aural-visual forms of the televisual discourse' (Hall, 1973). And it is this discourse, which shapes the form via the language of signification as a necessary stage of the communication process. Therefore, for Hall, 'the event must become a 'story' before it can become a communicative event' (Hall, 1973).

For Hall the slippage between the message that is encoded and that which is received once decoded is a result of the mismatch or 'structural differences' between broadcaster and audience, 'encoder' and 'decoder'. He indicates that 'noise' – 'distortions', interferences and misinterpretations – 'arise from... the lack of equivalence between the two sides of the communicative exchange' (Hall, 1973,). However the key problem from a Marxist perspective is that the ideologies or 'codes' of the ruling classes have become 'naturalized' within the universal signification of the media; in effect concealing the signifying practices responsible for this. In other words by resembling what we already know and accept, media representations evade critique. In a sense the media's re-productions of events are 'encoded' with instructions of how they should be received or 'decoded'. However, these meanings are so universally accepted that the message – for example, people in Africa are starving – appears as source or what Lash (1994) refers to as the 'signal'; thus shared meanings have already been produced for the audience and are assimilated without question (Lash, 1994, 138).

61

Analogue noise (Encoding and Decoding)

If we take this at face value, then the net result is a 'literal' signifier, or that which no longer produces connotative meanings. Yet, in reality connotative meaning is not beamed subliminally into the minds of the audience but is often produced via the discursive practices and situated ideologies of the social world. Therefore, for Hall, audiences will always find unintended connotations in the process of 'decoding' signifiers. Likewise while the majority of an audience will loosely accept the hegemonic or 'preferred' meaning supplied by the producer, adapting slightly to fit their situated worldview (what Hall terms the 'negotiated reading'), a minority will reject the 'preferred reading' altogether and instead engender a counter-hegemonic or 'oppositional reading'.

Conclusion

In reviewing three theories of structuralist thought it is easy to see how they have been called into question in an era during which postmodern approaches to cultural texts have become the norm. Arguably, it is the development of structuralism into post-structuralism that has been more instructive. In addition to this, it can be observed that some of the more rigid perspectives on gender are incompatible with feminist thinking, which views with scepticism Enlightenment concepts of logic and reason. That said, the work of Lacan influenced French feminism and Strauss's bearing on Hall's work is particularly pronounced. Indeed, Hall's Encoding/Decoding is a foundational work within Cultural Studies, framing and shaping the way in which many text-based research projects have been carried out.

6. Poststructuralism

Introduction

In this chapter we look at the way in which three poststructuralist theories can be used to frame and shape the way in which we think about contemporary media, society and culture. The chapter begins with an overview of Jacque Derrida's model of difference from Of Grammatology (1968). The second section of the chapter turns to focus on the work of Roland Barthes and 'Death of the Author' (1977), which was originally published in 1967, and the notion that the meaing of a text is inscribed by its reader. Finally, in the concluding part of the chapter we turn to the work of Michel Foucault in The Order of Things (1970) in which he argues that 'truth' is not to be found in relationship between the text and the knowledge structures behind it but our contingent acceptance of these arrangements. First, however, we turn to what is actually meant by the term poststructuralism.

Poststucturalism

The primary difference between the structuralism and poststructuralism is that whereas, for the former, meaning is defined by the similarities between things, for the latter, meaning is constructed, or rather deconstructed via difference. Structuralism is essentialist, in the sense that it looks for deep structures that reflect the essence of humanity in its analysis of cultural texts. Poststructuralism, however, foregrounds linguistic meaning in situated human practices and discursive contexts. In other words, meanings change according to the form in which they are said and the cultural context in which they are interpreted. This has major implications for semiotic analysis in particular the way in which we interpret cultural commodities or 'texts'. From a poststructuralist perspective meaning is not inherent in a cultural text as an absolute, but is instead produced contingently in the interaction between subject and object, the observer and the observed. Besides its rejection of the grand narratives and deep structures of its namesake movement, poststructuralism, like postmodernism, which we look at in the next chapter, also rejects the idea of a consistent, centred and homogeneous human subject.

6.1 Jacques Derrida – Deconstruction and Différance

NAME: Jacques Derrida (1930 – 2004)

KEY IDEA: The structural components of a cultural text cannot be defined in relation to those of another, either on the grounds of commonality or binary opposition. Instead, Derrida views the structure of a text as multi-accentual: a snapshot of an endlessly shifting diachronic process in which each narrative has several different points of origin and destination.

KEY TEXT: *Of Grammatology* (1998).

Unlike the structuralists before him, Derrida does not seek grand narratives or an ontological position in his philosophical approach, but instead merely exposes the unseen in the philosophical works of others. In opposition to the ideas of Strauss, Derrida argues that structural components of a cultural text cannot be defined in relation to those of another, either on the grounds of commonality or binary opposition. In other words he is not deconstructing the text from the structuralist perspective of analysing structures synchronically - comparing parallel components of a particular structure, for example myth, in order to uncover commonalities - but instead views the structure of a text as multi-accentual; a snapshot of an endlessly shifting diachronic process in which each narrative has several different points of origin and destination. Thus Derrida argues that once these alternative dimensions are taken into account, grand narratives become distorted, revealing their ultimately dualistic or schizophrenic nature. Meaning, rather than being fixed and absolute, becomes nebulous: the signified decouples from the signifier and becomes but one of many alternative signifiers; and therefore meaning is no longer fixed but deferred, from one signifier to the next.

It is this notion of deferral that is central to Derrida's concept of différance. The term is derived from the French word différer meaning both to differ and to defer and is used to illustrate the way in which the meaning of a signifier is postponed as it will always refer to another in an endless chain of signification. A simple example of the way meaning shifts is the simple act of right clicking on a computer and using the word processing program that we have used to write these words to come up with alternative (substitute, other, unusual, different) meanings! Therefore the term différance is dualistic, meaning both deferral and difference. The second meaning, that of difference, is crucial to the way in which Derrida defines the signifier (a word for example) as a function of how it differs from alternative meanings. Derrida coined the term différance - deliberately misspelled from its origin in différer - to emphasis the slippage between written and spoken forms. He argues that this serves to remind us 'that, contrary to an enormous prejudice, there is no phonetic writing... What is called phonetic writing can only function... by incorporating nonphonetic "signs" (punctuation, spacing etc.)' (Derrida, 1998, 387).

The film industry is an excellent example of the way in which the meaning of a text is continually in flux. In part this can be attributed to the collaborative nature of film making, which in spite of celebrated example of directors who can be considered 'auteurs' still remains a very collective enterprise. However, it is also a function of the perpetuity of the medium. More so than other media forms, films are durable in the legacy they leave: framing and shaping the way in which whole generations understand and appreciate historical events, social issues and philosophical dilemmas. Perhaps only in pop music do we see cultural artefacts that endure in quite the same way. Advertising, television production and radio is, for the large part entirely contemporaneous and with few exceptions it is difficult to conceptualise the way audiences will perceive a text in the future. By contrast, film is imbued with a historical pertinence that hangs heavy over the work of actors, screenwriters and directors.

While film history is littered with commercial flops and productions that have met with hostile critical reactions, it is also an industry with a selective memory and propensity toward revisionism. Testimony to this is the number of films that have courted controversy upon release only to be rehabilitated years later. For example, MGM's Freaks is a film that, at the time of its release, was viewed as shocking: in part because of the way in which it depicted actors with physical disability but also because the film encourages to audience to sympathize with the perpetrators of a crime. Eighty years on and the film is regarded as a classic amongst early talking pictures, in part because the representation of physical disability is so very forward thinking and sympathetic. This is exemplary of the diachronic nature of symbolic meaning, which in Derrida's logic film has several points of origin and destination. And, indeed, the difference between its contemporary meaning and our understanding of the film today is the reason why it is a classic case study in issues of film and censorship.

Vivre la différence (Deconstruction and différance)

The same slippage can be observed in the way in which countless other films have been appropriated into contemporary popular culture. Stanley Kubrick's Clockwork Orange, for example, was infamous for the director's decision to withdraw the film in Britain. Though this decision was influenced by the advise of Kubrick's local police, who feared for the director's safety, in the popular imagination the title is often remembered as having been 'banned' by the British Board of Film Classification. This is of course incorrect and indeed the BBFC does not have the power to ban films. However, the controversy surrounding its initial release and the ambiguity of its less encumbered distribution in the US and Europe created a mystique surrounding the film until its eventual re-release after Kubrick's death in 1999. That upon second viewing the ability of the film to shock had diminished diachronically was offset by the enduring quality of the cinematography, acting and narrative. The contemporary perception of what Clockwork Orange symbolised is in this sense a contingent viewpoint coalesced around several points of origin and an ever evolving destination that is the present.

6.2 Roland Barthes – Death of the Author

NAME: Roland Barthes (1915-1980)

KEY IDEA: Influenced by Saussure's semiotics Barthes argues that the meaning of a text is inscribed by the audience who essentially re-write it. Barthes argues that in effect the reader becomes an author, rendering the providential creator of the text as good as dead: hence the title the 'Death of the Author'.

KEY TEXT: *Image-Music-Text* (1977).

Unlike Derrida, Barthes did not initially oppose the structuralist approach, but instead expanded it to incorporate the analysis of cultural products, which he viewed as the structural texts of modern capitalist societies. However, in his seminal essay 'The Death of the Author', he ceases to view texts as the products of their creators, to be analysed for the structural messages encoded in their production. Conversely, he suggests that the author falls victim to the neutrality of shared verbal discourse and dissolves from view.

Barthes argues that the notion of authorship is relatively modern and in some cultures the role of the narrator is to mediate between 'narrative code' and audience. Moreover, he claims that the status of the author is a direct product of modern 'capitalist ideology'. Indeed, he also suggests that it is this status, which generates the illusion of the 'author 'confiding' in us' (Barthes, 1977,142-3). Further to this Barthes argues that a text cannot be documentation or even representation of reality (i.e. autobiography) but instead dissolves reality and replaces it with a plateau of neutrality:

> [A] text is not a line of words releasing a single 'theological' meaning (the 'message' of the Author-God) but a multi-dimensional space in which a variety of writings, none of them original, blend and clash. The text is a tissue of quotations drawn from innumerable centres of culture' (Barthes, 1977,146).

Throughout his essay Barthes discusses the written 'text', however, the meaning extends to all cultural products and significations. Barthes argues that the meaning of a text is inscribed by the audience, who essentially re-write it. He suggests that in effect the reader becomes an author, rendering the providential creator of the text as good as dead: hence the title the 'Death of the Author'.

Inscribed by the audience (Death of the Author)

Barthes' influence can of course be felt on the work of Stuart Hall's Encoding/Decoding (1973) in which he argues that the audience negotiates the meaning of text in its reception. However, from a contemporary perspective, Barthes work also foretells of the ways in which the interactive possibilities of new media have empowered audiences. The world of animation production is a classic example of a media industry that flourished on the back of technological advance. Though animation can be traced back to the Nineteenth Century with flipbooks and zoetropes, its history as an industry is entwined with the film industry. Georges Méliès's accidental discovery of stop start animation, for example, lead to the development of many narrative devices that are now a standard part of post-production, including time-lapses, dissolves and multipal exposures. Just as digital technology has revolutionised moving image cinematography, so too has the use of computers revolutionised animation. Software packages like GIF Movie Gear allow audiences to produce their own animation for broadcast on YouTube. From a theoretical perspective it is clear that animation has been blurring the distinction between the real and the simulated for much longer than the term postmodernism has been in use. However, the proliferation of Computer Generated Imagery (CGI) and the convergence of animation with special effects in mainstream film have taken this to new levels. In the US, for example, the ABC soaps All My Children and General Hospital routinely use CGI in the depiction of supposedly realist narratives. On a domestic level, CGI is a common form of audience creativity in the digital age. As with much web 2.0 content these amateur productions serve to reinforce many existing conventions. However, curious hybrids have emerged: the Anime music video, for example, is a YouTube favourite.

From Barthes point of view, animation supposes authorial creativity on the part of the audience in the form of the willing suspension of disbelief. In this respect advances in CGI have to some extent restricted the multiple interpretations of animated texts: Eadweard Muybridge's Horse in Motion, for example, requires more creativity on the part of the audience than George Lucas's Star Wars franchise. As the lukewarm reception for Peter Jackson's CGI heavy remake of Earnest Schoedsack's King Kong (1933) suggests, for some this creativity is a key to audience pleasure. What those film makers that seek to use CGI as means of creating cinema that is realistic sometimes forget is that animation is a discrete sub-culture of the film industry with its own history. Pixar films, for example, have a tradition of inter-textual self-referencing in the depiction of objects and characters that recur throughout their films.

Barthes view of media texts is incredibly prescient, anticipating trends in the delivery of media content forty years in advance. This is particularly evident in the transformations that have taken place in television production in recent years, which has seen the concept of broadcast television replaced by that of narrowcasting. The schedule has been replaced by TV-on-demand and multi-channelling has seen the erosion of old school tensions between the paternalistic sensibility of BBC and the populism of Independent Television. Likewise, interactivity (pressing the red button) and the convergence of Internet, digital radio, and mobile phones mean that audiences command the means of consumption. Though YouTube and Reality Television pertain to place the audience at the centre of television production in many respects, they reinforce its long-standing conventions. The narrative structure of television shows like Channel 4's Big Brother are managed by the intervention of the production team in everything from the selection of contestants, the design of the house to the editing and selection of footage for transmission. In this respect, the show has as much in common with the improvisation based drama of Mike Leigh as it does the production values of documentary. Likewise, performance art on YouTube is very mimetic: Chris Crocker's diary entries for example rework the confessional mode of reality television. For Barthes the interactivity that characterises the behaviour of television viewers in the UK is exemplary of the way in which audiences ascribe meaning to a text. More so than ever, viewers are now the authors of their own viewing schedule while YouTube and MySpace facilitate more 'hands-on' interaction.

6.3 Michel Foucault – The Order of Things

NAME: Michel Foucault

KEY IDEA: All cultural products, including those of the mass media, are shaped by the accepted structures of knowledge within the cultures in which they are produced. However, unlike the structuralists, Foucault does not seek to uncover the 'real' meaning of a text by uncovering these structures. Instead he attempts to show how 'truth' is not to be found in relationship between the text and the knowledge structures behind it but is contingent on our acceptance of these discourses as the foundation of truth.

KEY TEXT: The Order of Things: An Archaeology of the Human Sciences (1970).

Like Barthes, Michel Foucault was another philosopher to disassociate himself from structuralism and to break away from the existentialist trend that followed in the wake of Sartre. Foucault has been described as a cultural archaeologist because he studies social structures from the past, by uncovering them in the cultural texts of their time. In particular he looks at power structures and the way that they are manifest in the social institutions and cultural texts of the past (see chapter 1). However, it is Foucault's epistemological work – the study of what we know, how we know what we know and why we accept that which we know – which is most useful in interpreting media texts. For Foucault all cultural products, including those of the mass media, are shaped by the accepted structures of knowledge within the cultures in which they are produced. However, unlike the Structuralists, Foucault does not seek to uncover the 'real' meaning of a text and by uncovering these structures. Instead he attempts to show how 'truth' is not to be found in the relationship between the text and the knowledge structures behind it but is contingent on our acceptance of these discourses as the foundation of truth.

In 'The Order of Things' (1970), Foucault questions the fixed and deterministic nature of knowledge structures. He challenges the Straussian perspective that the meaning of a structure is determined by the relationships between the things within it. Instead, he argues that meaning is only determined by the discourses through which we categorise the structural relationship between, or 'order' of, objects ('things') in the external world. One such discourse is that of science. In Foucault's logic science is not the 'truth' but is instead a snapshot of 'knowledge' that is produced within a culturally specific discourse (Western science), and at a particular moment in time.

The central argument of 'The Order of Things' is that the conditions of discourses like the sciences are subject to diachronic change. Moreover, Foucault suggests that discourses through which we determine something to be 'acceptable' knowledge – what he terms episteme – are culturally, geographically and temporally situated. In particular he argues that there has been a radical shift in scientific thought, from a positivist system of taxonomic ordering of pre-modern or classical society – from the end of the Renaissance era to the middle of the Eighteenth Century – to a system of reflexive interpretation; both contingent and relativist in its abandonment of 'the space of representation' (looking out upon and making sense of things), in favour of an externalised view of mankind, and his works in progress (Foucault, 1970). Thus Foucault controversially argues that 'man is only a recent invention… and will disappear again as soon as… knowledge has discovered a new form' (Foucault, 1970).

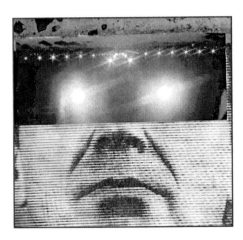

Knowledge is power (The Order of Things)

A clear example of how knowledge structures alter the depiction of 'reality' is the representations of climate change science in the media. In particular the way in which cultural differences and commercial and political interests shape the production of knowledge. And how these differing episteme, in turn, shape representations of scientific uncertainty in relation to the climate change. It is undeniable that the mass media is primary source of information on scientific developments. However, scientific uncertainty is rarely represented despite the fact that scientists themselves continually emphasize the conditionality of statements to highlight their awareness of the scientific uncertainty of their work. It would seem that the scientific tradition of reflexive relativism is at odds the mass media's tradition of positing knowledge as a pre-discursive polemic. In other words the "wishy-washy" reality of scientific discourse will neither corroborate opinions nor sell newspapers.

Since the notion of man-made climate change first entered the media in the UK in an article in the Saturday Evening Post entitled 'Is the World Getting Warmer?' (1950), there has been an ideological division between the medias of Britain and America in the way in which they have chosen to represent the issue. Moreover, cultural and economic differences between the two countries have also had a fundamental influence on the way in which scientific knowledge is both produced and represented. We have previously stated the importance of the media as a source of knowledge. However, this is not a problem in itself. The key issue regarding media representations of "knowledge" is the influence of the media conglomerates and commercial and political forces; who all have vested interests in representing one or the other side of the climate change argument. Furthermore, both countries have cultural traditions that affect attitudes to climate change, which arguably influence the way in which knowledge on the subject is produced across a whole culture. One the one hand, in the UK attitudes to national heritage, traditional institutional traditions and a more compact physical geography to have arguably influenced attitudes to environmental conservation. On the other hand however, protestant traditions and a national ethos of economic freedom with higher levels of personal consumption patterns in the US have swayed the national bias towards climate scepticism.

Conclusion

In reviewing three poststructuralist perspectives it is easy to see how influential they have been on the development of media and cultural theory, offering fascinating insights into the way in which contemporary texts make meaning. Most interesting perhaps, is the tension that exists between the notion of the meaning of the text as fixed, and the idea that it might be somewhat unstable. In particular the notion that a text's meaning is inscribed by the reader, not only reflects the way in which contemporary audiences engage with media but also draws upon de Saussure's view of the relationship between the signifier and signified as arbitrary. Moreover, the notion that the meaning of the text is continually in flux chimes with the work of Volosinov, who considers the meaning of signs to alter diachronically. Post-structuralism is, in this sense, particularly useful because not only does it pre-empt the shift towards postmodern cultural forms in the second half of the Twentieth Century but also because it configures these changes with a lineage of semantic meaning.

7. Postmodernity and Consumer Culture

Introduction

In this chapter we look at the way in which three theories of the postmodern can be used to frame and shape the way in which we think about contemporary media, society and culture. The chapter begins with an overview of Jean Baudrillard's analysis of consumer culture in the Consumer Society (1970); drawing upon the work of Marx and Saussure, Baudrillard argues that with the proliferation of information technology symbolic value has not only supplanted use value in capital acts of material exchange but that consumption is indivisable from the signification of personal identity. The second section of the chapter turns to focus on Pierre Bourdieu and his work on the commoditisation of taste in Distinction (1979). In his logic, the structure of society is shaped by cultural capital: the knowledge resources pre-requisite to engaging in the sign system. Finally, in the concluding part of the chapter we focus on the work of Frederic Jameson and his book Postmodernism, or, the Cultural Logic of Late Capitalism (1984). In general terms Jameson's work draws upon the work of Baudrillard and Bourdieu in his emphasis on the commoditisation of culture, the proliferation of information technology and the collapse of the distinction between the real and the simulated. However, for the purposes of this chapter we will focus specifically upon the terms parody and pastiche, which he uses to determine two dominant yet distinct modes of postmodern cultural activity. Firstly, however, we will turn our attention to what is actually meant by the term postmodernism.

7.1 Postmodernism – after modernism

In a literal sense postmodernism has to be understood in relation to modernism. While the former is characterised by playful uncertainty and ambiguous meanings, modernism sought to uncover new truth through questioning and self-conscious experimentation. Though modernism stood in opposition to the subjectivity and tradition of romantic thinking, it invoked an alternative set of meanings that in truth were no less arbitrary than that which preceded it. For example, in visual arts, modernist trends like abstraction and cubism may have reframed the relationship between the medium and that which is depicted (the signifier and the signified), however they still operated within fairly stable conceptions of what art is, and the role of the artist. Postmodernism sought to break down this fourth wall and encourage cultural producers to re-think the relationship between text and audience. In this sense postmodernism could be said to invoke aspects of semiotics that can be traced back to the Nineteenth Century: the arbitrary relationship between the signifier and the signified. And, indeed the term postmodern has been in use since the 1870s. However, there is a distinction between its use as an adjective and a proper noun to define a specific epoch.

The use of the term Postmodernity to identify an era in late-capitalism is the basis upon which Baudrillard, Bourdieu and Jameson are writing. Generally speaking that period is seen as beginning after the Second World War in the 1950s and is characterised by the proliferation of consumer lifestyle, credit and technology. Its apex could be said to embody the social changes the occurred in Western society during the 1960s, with the rise of Feminism, decline of the church and the emergence of a more relaxed attitude towards issues of sex, class and race. Laid over this of course has been the shift to a service economy and the proliferation of domestic communication technology blurring the boundaries between what is real and what is simulated. It is, in this sense, difficult to differentiate between that which is postmodern and that which is symptomatic of Postmodernity. Cinema, for example, which was popular long before the Second World War requires the suspension of disbelief and a willingness to accept as true simulated fictional environments: all characteristics of postmodern experience. Moreover, it is a commoditised cultural form utilising communication technology. However, it cannot be viewed as symptomatic of Postmodernity until much later in its history when audiences began to identify themselves as consumers and engage in the reflexive construction of their own personal identity; i.e. by going to see a particular film they are self-consciously assembling a narrative of the self. And indeed, this is the basis upon which Baudrillard considers the role of the consumer in postmodern culture.

7.2 Jean Baudrillard – The Consumer Society and Hyper-reality

Name: Jean Baudrillard (1929 to present)

KEY IDEA: The proliferation of information technology alienates man from real lived social existence, forcing him to enter a new media induced reality known as hyper-reality: hyper-reality is characterised by the collapse of the distinction between the real and the simulated and the predominance of the simulacrum.

KEY TEXT: *The Consumer Society: Myths and Structures* (1970).

Like Marx, in The Consumer Society (1970) Jean Baudrillard is pre-occupied with the economy. For Baudrillard, shifts from manufacturing to information-based industry are characteristic of the emergence of postmodern society. However, it is at the point which engagement with the economy becomes more tangible at the level of consumer than producer that is most important; this inaugurates a new sensibility whereby the consumption of goods is indivisible from signification of identity. Baudrillard calls this semiotic landscape 'hyper-reality'. While Baudrillard rejected the vision of this explored in Andy Wachowski 1999 film The Matrix, Peter Cattaneo's 1997 film The Full Monty arguably presents an account of how we got there: the decline of heavy industry and the collapse of traditional values and roles. For Baudrillard, ambience is also a key concept in postmodernity; this he believes is a function of a society in which mankind is alienated from each other:

> The concepts of 'environment' and 'ambience' have undoubtedly become fashionable only since we have come to live in less proximity to other human beings, in their presence and discourse, and more under the silent gaze of deceptive and obedient objects which continuously repeat the same discourse, that of our stupefied power, of our potential affluence and of our absence from one another (Baudrillard, 1970, 29)

The simulacrum (The Consumer Society)

Illustrative of this is the grouping of consumer products, not in relation to their use or function but according to their ambience. Exemplary of this is the well-known advertisement for French soft cheese which reads 'du vin, du pain, du Boursin'. Unified only by their transposition into a foreign language, which in itself connotes sophistication, the three items (bread, wine and cheese) create a potent new symbol of rustic French cuisine that is almost biblical in it simplicity. The importance of ambience for Baudrillard is predicted on the collapse of the relationship between the signifier and the signified, the real and the simulated and the emergence of a new sign: the simulacrum.

The production of the simulacrum, or the copy without an original, is one of the key theoretical issues in Popular Music Studies: in particular, the blurring of the distinction between the real and the simulated. Some argue that musical recordings are the epitome of the postmodern text. 'Records', for example, are not usually recorded live but artificially 'constructed' in the studio. The music video is a good example of this as they rely heavily upon inauthentic performances and abstract visuals: storylines are often very impressionistic and the artist is usually miming. Others thinkers have tried to identify key moments in the history of popular music when it seemed to embody postmodern cultural practice e.g. the advent of digital sampling or the proliferation of music video in the 1980s. The problem, it would seem, is that it is impossible to find a pre postmodern moment: from the gramophone to YouTube, popular music culture is inherently synthetic.

In Baudrillard's logic we have reached a point where the whole of modern life is commoditised in ways that are characteristic of the shopping mall and the modern airport:

> [A]ll activities are sequence in the same combinatorial mode; where the schedule of gratification is outlined in advance, one hour at a time; and where the 'environment' is complete, 'completely climatized, furnished, and culturalized. (Baudrillard, 1970, 33)

76

The problem with this according to Baudrillard is that human desire and aspiration is restricted to the desire to possess what other people have. Individualism is seen as in no way contradictory to the resembling everybody else. The reason for the success of consumer culture in Baudrillard's logic is twofold. In the first instance, consumerism offers the promise of total fulfilment. Secondly, consumer culture constitutes a new and authentic language. Moreover, consumerism is perceived as a harmless way in which the individual expresses his ego. That said, Baudrillard argues that consumerism is as meaningful as any other human interaction. However, perhaps where Baudrillard is most instructive is in his assertion that the relationship between the order of objects and human interaction:

> [O]bjects are categories of objects which quite tyrannically include categories of persons. They undertake the policing of social meanings, and the significations they engender are controlled. Their proliferation, simultaneously arbitrary and coherent, is the best vehicle for a social order, equally arbitrary and coherent, to materialize itself effectively under the sign of affluence. (Baudrillard, 1976, 413)

Within consumer society the notion of social status and social standing are in this sense one and the same. As Baudrillard states: 'there is not real responsibility within a Rolex watch' (Baudrillard, 1976, 415). As a code then the system of objects underpinning consumer culture may appear to be transparent; however, it conceals according to Baudrillard the real relations of production and social interaction. Consumer culture is in this sense a systematic manipulation of the sign system to suit the interests of the ruling class.

A clear example of this could be the way that companies like Reebok and Nike market premium brand sports wear. In the first instance, the purchase of branded sports items appeals to the individuals desire to be different: to purchase a Nike or Reebok product is to distinguish oneself from individuals sporting generic and non-branded items. And yet, these items are not exclusive. They are in fact mass-produced and the purchase of such an item does not make the individual more unique but more ordinary. The myth, however, that both Nike and Reebok are selling, is the myth of total fulfilment and the opportunity to participate in the dominant sign system of contemporary culture: branded consumer goods. This expression of ego in the form of purchasing sportswear is, however, relatively harmless and comes with minimal risk. In addition to this the meaning of the item is in dialogue with the social standing of the purchaser. For example, if the consumer is an accomplished sports person then the object confirms this. If the person is less able, then the item confers the potential for achievement: the virtual opportunity for success. In this sense the type of sportswear they possess determines the social position of the individual. The concealed international dimension underpinning the labour relations involved in the production of branded sportswear for multi-national corporations however, evidences that this code is less transparent than it might first appear. Many companies like Nike and Reebok locate factories in the developing world where production costs are cheaper and working conditions less heavily regulated. As Baudrillard suggests, however, this power dynamic is completely concealed in the semantics of Western consumer discourse. It is to the complex way in which commoditised cultural forms are valued in Western culture that we turn to next in our consideration of another French theorist Pierre Bourdieu.

7.3 Pierre Bourdieu – Education, Taste and Cultural Capital

NAME: Pierre Bourdieu (1930 to 2002)

KEY IDEA: Social class is constructed by cultural taste; cultural taste is produced by education. Social class facilitates access to education and so cultural order replicates itself. In the process of education, the individual acquires cultural capital, which gives the individual the ability to identify culturally noble activity. Culture evolves through the nomination of new cultural activity as noble by individuals who are highly educated in the process of naming.

KEY TEXT: *Distinction* (1979).

The relationship between cultural and symbolic consumption is well mapped in the trajectory of Bourdieu's work in Distinction (1979). Bourdieu argues that class determines cultural consumption and in this sense his work draws heavily on the work of Adorno. There is what he terms legitimate and illegitimate taste and this is indexed precisely to a person's education. Good taste is the preserve of people with 'the right kind of education' and it tends to reinforce what he refers to as the dominant ideology. Bourdieu identifies what he calls 'cultural nobility' (Bourdieu, 1979, 2) in the claims that are made by and on behalf of divergent cultural matter. In his logic cultural taste is the site of contestation between differing accounts of what is legitimate and 'noble' and what is not. This he argues is contingent not upon the objects themselves but the way in which they are objectified. For Bourdieu, education is not the key to legitimacy itself but the successful proclamation of that.

The definition of cultural nobility is the stake in a struggle which has gone on unceasingly, from the seventeenth century to the present day, between groups differing in their ideas of culture and of the legitimate relation to culture and to works of art, and therefore differing in the conditions of acquisition of which these dispositions are the product. Even in the classroom, the dominant definition of the legitimate way of appropriating culture and art favors those who have had early access to legitimate culture, in a cultured household, outside of scholastic disciplines. (Bourdieu, 1979, 2)

The inextricable relationship, however, between the right sort of education and the propensity to make the right sort of claims about culture renders the individual somewhat impassive. Education and the successful naming of cultural matter as 'noble' are not mutually exclusive. Far from it, the sort of claim an individual makes is closely tied to the education they have had. In this sense, 'nobility' has the tendency to always foreground itself.

One of the ways in which Bourdieu's work is most useful, is the equivalence he perceives between the exertion of taste and discourses in consumption. For Bourdieu, cultural nobility is always negotiated in the 'consumption' of cultural matter:

Consumption is, in this case, a stage in the process of communication, that is, an act of deciphering, decoding, which presupposes practical or explicit mastery of a cipher or code' (Bourdieu, 1979, 3).

He uses the term consumption both literally and metaphorically to regulate what he refers to as cultural capital. On the one hand, cultural capital determines the predilection of consumers in the choices they make. On the other hand, all sorts of symbolic consumption goes on in the way knowledge is socially distilled. He gives the example of the value difference between a 'Concerto for the Left Hand' and a 'Straus Waltz' and quite clearly we can apply the same principal to other hierarchical relationships. Classical music versus pop music, rugby versus football, television dramas versus game shows are all binaries connected with the social standing of the audience. And, of course, this is not restricted only to class: sex, race, age, national and regional identities all influence the ability of the individual to make proclamations of cultural nobility.

A sign of good taste (Distinction)

The music press operates in just this way. Magazines likes Q and Kerrang guide their readers as to what music products they should purchase; they institute a framework of consumer based values for judging aesthetic works. Central to this is the education of their reader in the process of naming and the specialised world of rock history and canon formation. There is, however, a subtle difference in the weighting of education and judgment between different titles. On the one hand, Kerrang pertains to invite readers to make their own judgments of taste, while simultaneously educating its readership with subject matter that is on the brink of the mainstream. On the other hand, Q offers artefacts that are more familiar but makes an explicit attempt to educate its readers in the process of naming. This distinction is very revealing about the audiences for each magazine. While Kerrang has a mixed demographic of younger male and female consumers, Q is predominantly read by an older group of male readers. The slippage then between education and taste is connotative then of wider narratives of cultural power connected to the gender and age of the audience. In this sense readers of Kerrang can be seen as more receptive to new cultural forms because their own subject position is less fixed. By contrast, Q tends to reinforce the hegemony of white middle class patriarchal society. It is to the politics of cultural consumption that we turn to in the proceeding section, which focuses on the work of Frederic Jameson.

7.4 Frederic Jameson – Postmodern parody and postmodern pastiche

NAME: Fredric Jameson (1934 – present)

KEY IDEA: Building on the work of Baudrillard, Jameson argues that the distinction between the real and the simulated becomes very blurred in postmodern society. He uses the terms parody and pastiche to explain the way people use and borrow existing cultural artefacts. Pastiche is basic mimicry, while parody is more knowing and ironic.

KEY TEXT: *Postmodernism or The Cultural Logic of Late Capitalism* (1991).

Like Bourdieu, Frederic Jameson's account of contemporary consumption is inextricable from Frankfurt School anxiety about 'high' and 'low' culture. In 'The Politics of Theory', for example, he makes problematic the distinction between folk art and mass culture:

> The older kinds of folk and genuinely 'popular' culture which flourished when the older social classes of a peasantry and an urban artisanant still existed and which, from the mid-nineteenth century on, have gradually been colonized and extinguished by commodification and the market system. (Jameson, 1988, 112)

Like Baudrillard, Jameson perceives the death of a fixed and autonomous subject to be a function of a fundamental break in what he terms 'the cultural logic of late capitalism'. However, where his account differs is his insistence that this can be theorised from a neo-Marxist framework. For Jameson there is equivalence between the transformation from use to exchange value and the changes manifest in the economic culture of the 1980s. Central to this is the assumption that, like Baudrillard's model of the simulacrum, exchange value is inextricable from identity discourse: 'In contemporary terminology, then, we might say that 'use value' is the realm of difference and differentiation as such whereas 'exchange value' will as we shall see, come to be described as the realm of identities' (Jameson, 1991, 221). This is significant because it configures Jameson's model of contemporary consumption squarely with the appropriation of Marxist terminology by the Frankfurt School.

Hyper-reality (Postmodernism)

In Postmodernism (1984) Frederic Jameson argues that in late capitalist society the distinction between the real and simulated becomes very blurred. A map, for example, of Albert Square from the BBC television soap opera Eastenders is no less real than a map of the Outer Hebrides if the person looking at the map has been to neither. And, indeed, the map of Albert Square might seem a considerable deal more real if the person has watched that programme on television. In this sense Jameson is clearly building on the work of Baudrillard, but arguing that the simulacrum is simply a generalisable feature of late-capitalist society. And, to this end it could be argued that Jameson's work is interesting because it does not preclude the possibility of the revolution delineated by Marx and Engels in The Communist Manifesto. However, where Jameson's work is extremely useful is in his use of the terms parody and pastiche. He uses these to distinguish between the way the different ways in which people use, borrow and re-work cultural artefacts:

> Pastiche is, like parody, the imitation of a peculiar or unique, idiosyncratic style, the wearing of a linguistic mask, speech in a dead language. But it is a neutral practice of such mimicry, without any of parody's ulterior motives, amputated of the satiric impulse, devoid of laughter and of any conviction that alongside the abnormal tongue you have momentarily borrowed, some healthy linguistic normality still exits. Pastiche is thus blank parody, a statue with blind eyeballs (Jameson, 1991, 17).

To pastiche something is to mimic without any satiric or other impulse that communicates difference, but to parody is far more carnivalesque and knowing. Carnivalesque in this sense can be linked to the work Mikhail Bakhtin (1964) discussed in chapter 1, whereby the inversion of dominant cultural forms and inversion of traditional hierarchies generate agency for the individual.

An example of this can bee seen in the work of pop star Robbie Williams. On the one hand, the video for the song 'Millennium' (2000) is self-conscious in the way in which it plays with the iconography of James Bond. It is a tribute to 007 but plays with the difference between the ex-boy band member and Ian Fleming's suave secret agent. It is a homage to Bond but also a send up of his own contrived persona and is consequently pure parody. On the other hand, his Swing When You're Winning (2003) album betrays a quite different aesthetic rationale. Superficially it adheres to quite similar conventions – reworking a historical period – in this case rat pack era Frank Sinatra and Dean Martin. However, where Millennium is carnivalesque, Swing When Your Winning is laboured. Williams own vocal style is a combination of cabaret croon and cruise ship swing and the arrogance in tackling this particular American songbook is tangible. No amount of 'look at me mum' banter with the audience can conceal that this is not only an artist hopelessly out this depth but offering a very weak pastiche of a by-gone era. What we can extract from this model, however, is that the symbolic value of cultural material not only shifts diachronically but also depends on what is being revisited and how that matter is re-appropriated. In the case of Robbie the camp nature of James Bond perhaps lends itself to parody, while the more serious affectations of Sinatra et al are less easy to caricature. Not all recycling is equal: a plastic bag made of reprocessed petro-chemicals is more acceptable than a car made of scrap metal. However, it is the process of recovery that is perhaps the most important in determining the cultural currency that is carried.

Conclusion

In reviewing three postmodern approaches to popular culture, it is easy to see how they have framed the way in which commoditised cultural forms are understood. Most interesting perhaps, is the tension that exists between capitalism and consumerism: the way in which individual engagement with the economy is more tangible at the level of consumer rather than producer. In part this is because this conflict is the backdrop against which the proliferation of information technologies has taken place. Likewise, the oblique connexion between the symbolic meaning of commoditised cultural forms and the underlying power structures of capitalist society is a cause for concern. It is perhaps unsurprising therefore, given the fragmented nature of consumer culture and the collapse of the distinction between the real and the simulated, that instances of cultural resistance often take the form of carnivalesque mimicry rather than traditional notions of political expression.

8. Consumer Agency and Cultural Studies

Introduction

In this chapter we look at the way in which cultures of consumption can be deregulated from Marxist analysis. Instead we focus on the way style and taste can be used in the negotiation of personal identity, political agency and the construction of cultural history. The chapter begins with an overview of Dick Hebdige's work Subculture: The Meaning of Style (1979) with specific reference to punk, the safety pin and the reflexive construction of self. Hebdige's work can be seen to build upon the work of Jean Baudrillard and Frederic Jameson discussed in the preceding chapter. The second section of the chapter then turns to focus on Stuart Ewen, who's work is directly influenced by that of Hebdige. Ewen explores the endless recycling of the culture industry and the way in which creative output becomes cultural waste matter. Finally, in the concluding part of the chapter we focus on the work of Chris Anderson who explores the way in which the Internet and digital technology has transformed the economic structure of consumer culture. In particular his analysis of the 'long tail', niche marketing and the notion that the future of business is selling more of less has opened opportunities for consumer agency. Agency in this sense is perhaps best understood as virtual power and potential for action. Firstly however, we want to focus on the term consumerism in more detail. With specific reference to the work of Philip Woods we will to consider the ways in which spheres of culture superficially disconnected from capitalist acts of material exchange can be reconfigured as commoditised forms.

8.1 Defining Consumerism

In a simple sense consumerism can be understood as the belief that the purchase of material goods and professional services will result in psychological happiness, personal fulfilment and social regard. An early critique of this perspective came from the Norwegian-American Sociologist Thorstein Veblen. In The Theory of the Leisure Class (1899) Veblen argues that it is man's instinct to emulate and impress through the conspicuous consumption of material goods; for him it is this animal impulse that underpins the success of the capitalist system. This is an idea that Vance Packard's draws upon in The Waste Makers (1961). In this book, Packard expounds the sociology of capitalism for a popular audience and brings concepts like 'built in obsolescence' into public conscience. However, for the purposes of this chapter I want to define consumer culture in relation to an unlikely source: Philip Woods work on British education published in 'Parents As Consumer Citizens' (1993).

'Parents as Consumer Citizens' is an extremely useful way of beginning to think about consumer culture because it offers four clear ways in which consumer culture operates. In essence it is an inversion of Distinction (1979), in which Bourdieu argues education foregrounds taste. In 'Parents As Consumer Citizens' Philip Woods looks at the way taste foregrounds education in the multiple ways in which parents consume their children's education. The article begins with an outline of how reforms in British state education in the 1980s were designed to address parents as consumers of their children's education. He then offers four models of consumer engagement, each underpinned, like Baudrillard, by the notion that consumption is active and not passive:

1. Competitive Market Model

Schools offer a competitive market model of customer choice because information contained in league tables enables parents to assess schools on their past performance.

2. Personal Control Model

Like DIY furniture parents can also do things to influence the quality of their child's education, from help with homework to involvement in the PTA.

3. Quality Assurance Model

Parents are also empowered by the outlining of standards and specifications about the 'goods and services' they can expect in both the National Curriculum and school policy documentation.

4. Participative Model

Parents can engage in dialogue with providers of their children's education in the form of parents' evenings and other meetings with staff.

This can be transposed into multiple sphere in which the public can be seen to engage with facilities and services under state jurisdiction including the police, NHS and utility services. However, it can also be extended to socio-economic arenas not specifically connected with the consumption of state services or capitalist acts of material exchange.

To fully understand how Wood's model of consumer agency works, lets explore it in relation to two spheres of cultural traditionally dislocated from consumer narratives. In the first instance, the seaside resort offers a competitive market model of customer choice because visitors are able to choose between different resorts at home and overseas. Secondly, visitors have control over how they choose to use the seaside resort: what activities they engage in and what attraction they visit. Thirdly, local authorities within a coastal area also have to make sure that national standards of seawater quality and environmental cleanliness are met. And fourthly, visitors can engage in dialogue with authorities by writing letters of praise and complaint. Likewise this model can be applied to the consumption of the popular press. In the first instance, the popular press offers a competitive market model of customer choice because consumers are able to choose between different newspapers: the Sun, the Mirror or the Daily Mail. Secondly, readers have control over how they choose to use the newspaper: this could include the order in which they read it, sections they omit and what they choose to do with it afterwards. Thirdly, the owners of the newspaper have to make sure that certain standards of journalism are met in terms of accuracy, balance and fairness of representation. And fourthly, readers can engage in dialogue with editors, individual journalists or the Press Complaints Commission.

Testimony to the power of consumer ideology is the way in which successive governments have instituted target driven, quality-assured models of state policy to engage the electorate. However, there is a difference between the self-interested consumption of disposable goods and the way in which individuals engage in political ideology. Indeed, it is arguable that the reforms of education, the NHS and British police force have delivered highly spurious improvements to the electorate. Before we take-on Wood's model wholesale then, it should be remembered that like many of Margaret Thatcher's reforms to the civil service in the 1980s, which appropriated consumer models of accountability, the remodelling of education during this period was as much about economic efficiency as empowerment of the electorate. For a more detailed discussion of hegemony and Thatcherism see chapter 4 and our analysis of the Italian neo-Marxist Antonio Gramsci. The key issue it would seem is that while the wilful squandering of the public purse is clearly unacceptable; the benefits of some expenditure clearly cannot always be measured empirically. For example, the benefits of spending on hospital food and school dinners cannot be quantified exactly. This is the tension then between capitalism and consumerism. While capitalism measures benefits in terms of fiscal expedience, consumerism takes account of the intangible, symbolic and creative ways in which the individual appropriates mass culture. Consumer culture is in this sense intimately linked to the way in which both Baudrillard and Jameson discuss the ideological function of material goods in late capitalist society. Although both theorists consider the innovative and figurative dimensions of consumer culture, their critiques also emphasise the role of consumerism as a function of bourgeois hegemony. However, for the purposes of this chapter, we will focus on three theorists who emphasise the potentially resistive qualities of consumer culture.

8.2 Dick Hebdige – Subculture and the Meaning of Style

NAME: Dick Hebdige (1951 – present)

KEY IDEA: Central to Hedge's view of consumer culture is the notion that the individual is active in the ascription of meaning to consumer goods. Focusing on the punk movement of the mid-1970s he looks at the way in which youth-cultures borrow and re-work the symbols of preceding youth groups. In particular his semiotic analysis of the safety pin as a symbol of cultural rebellion has been particularly influential in framing and shaping the way in which proceeding moments of cultural resistance have been understood.

KEY TEXT: *Subculture: The Meaning of Style* (1979).

The theme of resistance is explored in the seminal work by Dick Hebdige Subculture: The Meaning of Style (1979), which builds not only on the work of other cultural theorists from the Centre for Cultural Studies in Birmingham (particularly Paul Willis) but also the work of French sociologist Jean Baudrillard. Like Baudrillard, Hebdige sees the purchase of material goods as an active process in which the self is reflexively constructed in dialogue with the meaning of objects. Hebdige reframes this process, however, and focuses instead upon the way in which audiences imbue objects with ideological meaning. Using the example of the punk safety pin, he suggests that when appropriated by minority groups or 'subcultures' such objects take on a new semantic meaning.

Hebdige begins the book with a citation of the novel The Thief's Journey by Jean Genet, in which Spanish police interrogate the protagonist and proclaim his homosexuality by the presence of a tube of Vaseline. Hebdige claims that he is interested in this because it is symptomatic of the way in which mundane objects take on a symbolic dimension. It is, however, the tension between the object and the reaction that it elicits from an audience that makes it meaningful. In particular, Hebdige is fascinated with Genet's jubilation at the unambiguous nature of the way in which the officials interpret the tube of Vaseline: it is in this sense a semantic triumph for the otherwise marginalised and oppressed.

Citing the legacy of Duchamp, whose work challenges traditional conceptions of the relationship between artist and artwork, Hebdige turns his attention to the way in which punk appropriates manufactured objects from a variety of contexts to disrupt normative messages and values:

> Objects borrowed from the most sordid of contexts found a place in the punk's ensembles: lavatory chains were draped in graceful arcs across chests encased in plastic bin-liners. Safety pins were taken out of their domestic 'utility' context and worn as gruesome ornaments through cheek, ear or lip (Hebdige, 1979, 1067)

In particular he emphasises the symbolic resistance of items that include PVC, bin-liners, fake blood, rapist masks, leather bodices, and the paraphernalia of bondage. However, where Hebdige's work is transforming perhaps, is the notion that these strategies did not just disrupt the wardrobe but called into question the dominant ideological structure of mainstream culture.

For Hebdige the clothing styles of punk are seen as emblematic of an entirely separate set of social values. Overt displays of romantic courtship are for example eschewed amongst young persons identifying themselves as punks. Individuals who prefer to listen to atonal music that emphasises passion over technique likewise reject free form and ritualised forms of dancing. However, perhaps the key significance of punk for Hebdige is its connectedness to working class modes of expression. In this sense punk is significant because it negotiates a semantic space for modes of expression outside of dominant cultural forms. The coherence of this, according to Hebdige, is inherently contradictory, however. On the one hand, punk symbolised chaos and disorder: it purposefully challenges existing frameworks of culture, taste and ideological structure. On the other hand, it is highly ordered: the coherence with which it symbolises anarchy is a product of a tightly controlled system of meaning. Hebdige borrows the term 'homology' from structuralist theorist Levi Strauss to describe this regimented nature of punk style.

Decontextualised (Subculture)

While punk appropriated forbidden signifiers into a coherent system of meaning, the problem it creates for a semiotician is that nothing is fixed and sacred. Even seemingly fixed symbols of alienation like the safety pin speak in an abnormal tongue, to borrow a phrase from Jameson:

> We could go one further and say that even if the poverty was being parodied, the wit was undeniably barbed; that beneath the clownish make-up there lurked the unaccepted and disfigured face of capitalism; that beyond the horror circus antics a divided and unequal society was being eloquently condemned. (Hebdige. 1979, 1073)

Yet for all it resistive potential, this carnivalesque sensibility eludes meaning. In the appropriation of the swastika by some punks, for example, the implications are unclear. On the one hand, it serves as a symbolic rejoindner to Nazi-Germany, which had 'no future'. On the other hand, the invocation of far right sympathy is at odds with punk's anti-facist sensibility. In Hebidge's logic then the appropriation of the swastika is an example of how the signifer can be detached from the signified. There are in this sense echoes of both Saussure's ideas and those of Volosinov. However, where Hebddige is particularly instructive is in his assertion that in this instance the swastika is 'dumb'; synonymous with an unidentifiable miscellany of values that render its meaning unintelligible. And indeed, the meaning of the swastika pre-dates Nazi Germany: archeologist have traced it back to the neo-lithic period and this certainly gives credence to Hebddige's view. The same can be said dumbness can be observed in punk's invocation of working class politics. According to Hebdige this is 'abstract, disembodied and decontextualized' (Hebdige. 1979, 1073). In this sense, punk's revolutionary potential is neutered somewhat by its nebulous nature.

There is of course a clear link here to the politics of camp as defined by Susan Sontag (1964): a Foucauldian semiotic in which the multi-accentuality of the text is liberated as a means of challenging the hegemony of patriarchal culture. And, of course the parodic quality Hebdige identifies in punk is redolent of this. Sub-cultural readings of media texts often emphasize their resistive qualities. Both Lyn Thomas (2002) and Len Ang (1995), for example, explore the complex ways in which female audiences configure popular television shows into their own personal narratives of self. Irony in this sense is key part of the way in which the meaning of the text is decoded and shared with other fans. More populist examples of this would be films that deliberately spoof the narrative conventions of a specific genre like Scary Movie (2000), Austin Powers (1997) or Airplane! (1980). That said, the politics of parody is extremely complex. Minority groups for example are wont to re-appropriate terms that are used to oppress (e.g. 'nigger' or 'queer') and give them new positive meanings. The rehabilitation of the film Freaks (1931) on the art-house cinema circuit of the 1960s, for example, reframed narratives of disability and the term 'freak' in a more positive context. However, perhaps the most complex contemporary example of sub cultural resistance is the use of the term 'chav', to describe a group of people belonging to one of Britain's poorer socio economic groups.

Defining what is meant by the term 'chav' is difficult and clearly the origins of the term can be found in a range of dialect words used to describe a child or teenager. However, in contemporary parlance the term is used to define somebody who is socially disadvantaged and aggressive: the nature of that social disadvantage can take a variety of forms including social, educational, economic or physical. On the face of things, 'chavs' as a social group could, therefore, be recouped within Marxist notions of the Proletariat. However, a key feature of the 'chav' is their refusal to acknowledge this disadvantaged social position. Though 'chavs' may adopt an explicitly aggressive attitude towards those belonging to the hegemony of the ruling class including teachers, police officers and the middle-class, this is ill-defined and lacking 'homology'. In addition to this, 'chavs' do not adhere to traditional notions of working class pride, deference to high status groups or aspirations for self-improvement. Whether or not they embody the spirit of sub-cultural resistance then is open to interpretation.

A key feature of the use of the term 'chav' is that it is generally used as a pejorative term by someone who identifies him or herself as belonging to a higher socio economic grouping. 'Chavs' are not self-identified but rather they tend to be somewhat oblivious to their class position: undermining the ideal of self-determinism intrinsic to notions of consumer agency. Moreover, the term is seen as synonymous with social problems of an under-class that includes teenage pregnancy, drug abuse, petty crime and poor literacy. Indeed, numerous critiques have implied that it is a term used by middle-class snobs and that it is a form of social racism. In this sense, the term 'chav' could be viewed in neo-Marxist terms as an agent of hegemony: complicit in the acquiescence of social consent to the polarisation of wealth in the Noughties. However, one arena in which 'chavs' have exhibited behaviours redolent of what Hebdige defines as subculture is their appropriation of premium brand clothing including Burberry, Von Dutch and Helly Hansen. On the face of things this disrupts the uni-accentuality of brand meaning in a way that in Volosinov's terms would be symptomatic of class resistance. And, the reluctance of certain brands to be associated with this particular social grouping is symptomatic of a consumer agency of sorts. However, it is uncertain as to whether the appropriation of these designer labels actually is intended to subvert dominant hegemonic meaning. Indeed, it could be argued that the obliviousness of the 'chav' to their social disadvantage and the complicity of their aspiration with dominant value systems represents the inverse of class resistance. Likewise, the ease with which the 'chav' can be differentiated from bourgeois consumers of these brands reinforces class distinction. Indeed, it could be argued that the appropriation of premium brand products by people from lower socio economic grouping reveals the inverse of what Bourdieu's talks about in Distinction: as opposed to the 'capital of consecration' bestowed upon noble artefacts by embourgeoised consumers, the 'chav' carries a 'deficit of desecration' that renders previously noble cultural matter deficient and inferior. In the proceeding section we will discuss the relationship between the style politics of the counter-culture and its relationship with the capitalist mainstream.

8.3 Stuart Ewen – All Consuming Images

NAME: Stuart Ewen (1940 approx to present)

KEY IDEA: Style is political: visual signifiers encode systems of belief. While these visual codes are often long and their histories complex appropriation by consumer culture often dilutes their ideological potency. The ideological significance of the punk safety pin, example, is diminished when adopted by mass-produced clothing lines; what is left is in Ewen's terms 'cultural waste matter'.

KEY TEXT: *All Consuming Images: The Politics of Style in Contemporary Media* (1988).

The politics of style recurs as a key theme in the work of Stuart Ewen. In All Consuming Images (1987) he explores the way in which sub-cultural matter infiltrates the mainstream. He considers the way that subcultures change in terms of their meaning because of the media. Something that may have a long cultural history can be used up and spent rapidly when appropriated by a large capitalist enterprise. For Ewen what is left is simply cultural waste matter. Though Ewen is clearly influenced by Hebdige and Baudrillard, the shadow of Marx also hangs over his work in his explanation of the way in which mass-media exploits sub-cultural forms. However, it is perhaps his use of Hebdige's term 'bricolage' that is most instructive: in Ewen's logic this accounts not only for the improvisational nature of meaning attached to commoditised goods but the way in which waste matter can be recycled.

Ewen begins with the assertion that premeditated waste and changing style are the key features of a consumer society. In this sense he is drawing upon theories previously expounded upon by Adorno, Packard and Baudrillard. However, the significance of this for Ewen, is that in these changes we confront the passing of time:

> [A]s style becomes a rendition of social history, it silently and ineluctably transforms that history from a process of human conflict and motivations, an engagement with social interests and forces, into a market mechanism, a fashion show. (Ewen, 1987, 1082)

Ewen contends that until the 1950s this rendition of social history was fairly singular, depicting only the ideal cultural representation of the middle class. Later periods, however, he suggests are characterised by a more plural construction of social history cohered around what is nominally referred to as the 'alternative'. Alternative in this sense embraces oppositional concerns including issues of class and sexual oppression, political activism and global inequality. What concerns Ewen is what happens when this cultural archive is appropriated into the mainstream of consumer culture.

Ewen's view of the cultural crossover between alternative and mainstream is extremely cautious. Focusing on a number of instances in which material of the counter-culture has been exploited by capitalist institutions he suggests the end product is waste matter. For example, he looks at the use of the phrase 'right on', which had its origins in the black liberation movement of the 1960s; he discusses what happened when the term 'write on' was used by the BIC biro company in the 1970s:

> Whatever significance or value the expression may have had in the context of its earlier development, that value was now outweighed by its exchange value, its ability to make something marketable hip. When its marketability had been consumed, the phrase – like so much else – achieved the status of cultural waste matter. (Ewen, 1987, 1082)

93

In this sense Ewen's view of the culture industries echoes that of neo-Marxists like Adorno and Marcuse. However, where Ewen is particularly useful, is in the identification of a term used by Hebdige: bricolage. Bricolage as Ewen interprets it, is a term to describe the improvisational meanings that can be constructed from mass-produced goods. The process is, however, cyclical: while punk may re-work the meaning of items like the safety pin, so too do those alternative sets of meaning inform the mainstream. Mass-produced punk 'style' clothes, for example, can be found in high street department stores and their popularity ebbs and flows with the seasons. Though like all commercial forms these artefacts will eventually be consigned to the dustbin of history, they can, according to Ewen, be revived at any point for commercial recycling. This fluid view of the sign system echoes the work of Judith Butler in Gender Trouble (1990) discussed in the next chapter: meaning is in this sense continually in flux, for ever changing and contingent upon the context of signification.

Cultural waste matter (All Consuming Images)

Contemporary culture is of course littered with countless examples of the way in which cultural material from the past can be rehabilitated and appropriated by new generations of capitalist producers and consumers. Popular music culture is characterised by the recycling of its own history in the form of the cover song, re-release, compilation and comeback. The legacy of Swedish pop group Abba is exemplary of this trend. Relegated to the dustbin of pop for much of the 1980s, the group's rehabilitation began in 1992 with an EP of cover versions by the pop duo Erasure. A subsequent greatest hits album topped the charts and the groups place in the rock canon was reconsidered. Subsequent re-workings of their back-catalogue saw the launch of a stage musical Mamma Mia (1999) and cinematic adaptation of this in 2008. Indeed, in 2009, Mamma Mia the film was certified as the highest grossing film of all time in the UK. While the scale of Abba's enduring success is atypical, with the proliferation of dance music since the 1980s, subtle variations of this process have become commonplace. For example, it is now an everyday occurrence to listen to recordings composed entirely of samples. Typical of this is French producer Bob Sinclar's 2006 hit 'Rock This Party'; the track borrows heavily from a 1990 recording by C+C Music Factory entitled 'Gonna Make You Sweat (Everybody Dance Now)', which is in itself is a recording composed entirely of samples. In the short term, this transformation challenged traditional notions of authorship, however, as the industry has moved forward the skills of the d.j. and producer have been routinely integrated into dominant narratives. Likewise, the history of the film industry is characterised by re-makes, re-releases and the rehabilitation of films from the archives. Outside of media industry the same process can be observed in car-design. The launch of the 'new' Beetle, in 1998 by Volkswagen was part of a wave of fin-de-siecle excitement amongst car-designers intent on capitalising upon the nostalgia of consumer for cars of a previous era. Other products that exhibited the same trend include the 'new' Mini, Jaguar S Type and Rover 75.

In part the influence of bricolage upon contemporary patterns of consumption can be attributed to the accelerated nature of capitalist modes of production in the latter half of the Twentieth Century. While the first half of the Twentieth Century was characterised by ponderous mass production techniques, in its latter years improvements in communication between suppliers and retailers lead to a much more dynamic relationship between consumer and producer. The preferences and taste of shoppers began not only to influence stock control but also regulate production, making capitalism more effective and the opportunities for the reflexive construction of self more determined. In so far that the slickening of capitalist modes of production in the second half of the Twentieth Century was a product of developments in communication technologies including the telephone, fax machine and computer, the same technologies have also heightened our awareness of cultural history and influenced the direction of bricolage. Representations of social history through film, television and pop music serve as a rich archive for cultural production in both alternative and mass culture. And, indeed, the division between the two has become more blurred than ever before. Online retailers like Amazon, for example, may offer an homogenous brand identity, but are in fact an umbrella organisation for hundreds of thousands of independent retail outlets operating independently from domestic and backstreet premises. In the next section we will discuss the way in which the Internet has transformed the nature of consumer culture in the Twenty-First Century and the implications this has for the future of media production.

8.4 Chris Anderson – The Long Tail

NAME: Chris Anderson (1961- present)

KEY IDEA: The future of business is selling more of less. The Internet has revolutionised the distribution possibilities of capitalism. While business structures in the Twentieth Century were charaecterised by Fordist principals of mass production, businesses that have flourished in the Twenty-First Century are niche marketing focusing on a defined community of consumers.

KEY TEXT: *The Long Tail: How Endless Choice is Creating Unlimited Demand* (2006).

The fragmentation of the market is the backdrop for Chris Anderson's The Long Tail (2006) in which he argues that the future of business is selling more of less: inverting post-Fordist principles of globalised mass production. Anderson's key argument is that the niche marketing and distribution possibilities created by the Internet and other digital media have opened up the profitability of fringe creative industries. What Anderson describes as the 'long tail' is facilitated by the economics of virtual retail space. With no limit to the stock that can be carried, online retail – of music, for example – has seen a shift away from 'hits' towards fragmented niche markets and back catalogue: endless choice creating unlimited demand. (Anderson, 2006). The influence of this can be felt across all forms of contemporary media. For example, BBC 4 has an audience of only 1.8% of BBC 1 yet it commands 4.7% of the budget because its niche market productions can be sold other narrowcast TV channels around the world.

In this sense the significance of what Hebdige identifies as sub-cultures cannot be underestimated. With the proliferation of digital technologies and niche marketing, sub-cultural groups have become the key to profitability. For example, in the mid Noughties Kerrang became EMAP's new 'hero-brand', leaving behind bandstand titles like Q and Smash Hits, because it spoke to a more readily identifiable 'community of consumers'. Though its readership was lower than its EMAP siblings, the specialised nature of the magazine and its highly devoted readership fitted exactly the business model identified by Anderson. Of course, the profitability of such fringe publications is in part made possible because of the decreased cost of productions methods: desk top publishing, digital photography and file transfer software have reduced the overheads of independent operations. And, in this sense it could be argued that the past ten years has seen democratisation in access to the means of cultural production in the West. That this position of privilege is underpinned by the exploitation of workers in the developing world, who produce cheap electronic goods, is discussed in more detail in chapter 2 on Marxism. The corollary, however, of such an egalitarian view of creative endeavour in the West, is that contemporary media production assumes that audience will form their own improvisational meanings from mass-media forms. And indeed dialogue between producers and consumers is the key to Kerrang's success: by embracing interactivity in the form of MySpace, Facebook and the message board, the magazine is able to keep up with the shifting taste of its target audience.

This dialogic approach to the production of meaning informs contemporary ideas about advertising. The traditional view of the adverts as a paid one-way communication, in which the sponsor controls the message, has been replaced by texts that invite the audience to engage more creatively. For example, while slogans of the past may have instructed the audience that 'Guinness is Good For You' (Guinness 1930s) or to 'Go To Work On An Egg' (Egg Marketing Board 1960s) since the 1970s advertising has been more elliptical and cryptic. Likewise, humour has become a key weapon in the armoury of advertising creatives. In Salman Rushdie's copy for cream cakes ('Naughty But Nice'), Martini's 'anytime, anyplace, anywhere' campaign and Heineken's 'refreshes the part other beers cannot reach' advert, the use of innuendo draws upon the multiple meanings inscribed in the text by the audience. Clearly this reflects Barthes' contention that the meaning of a text is fixed, not at it moment of creation but in its reception.

Today advertising campaigns are increasingly interactive, operating across multiple platforms and utilising synergised brand allegiances. Vauxhall VXR's Sport Driver of the Year is a case in point: teaming up with men's mag FHM, the car company put together a competition for readers that invited them to partake in a series of race days that generated extensive coverage in four issues of the magazine. Such creative solutions to 'buying' media space is increasingly characteristic of agencies that promise to deliver a very specific audience to the client by strategically targeting specific demographic profiles. Advertising is increasingly orientated towards defined communities of consumers; what Hebdige would describe as 'sub-cultural groups'. Moreover, the resistive strategies deployed by such groups are often the key to some of the most effective forms of advertising. Small-scale guerrilla tactics for example can be more effective than high profile media campaigns. For example, Virgin Holidays suitcase campaign simply involved donating a branded Virgin suitcase to a series of well known civic statues: creating the appearance that the historical figures were going on a Virgin holiday. Such strategies reinforce Chris Anderson's mantra that the future of business is selling more of less and that niche market campaigns can be more profitable than the economics of mass production.

The same fluidity can be observed in the way in which journalism has developed over the last twenty years. The traditional view of the journalist is steeped in the objective accountability of the press as the Fourth Estate: framing political issues and offering political advocacy. For example, in 1991 the European Court of Human Rights ruled against the British government's use of a gagging order to restrain newspapers from reporting on former MI5 secret services officer Peter Wright's memoirs Spycatcher (1987). Likewise, investigative journalists work hard on behalf of the electorate to expose government scandal. For example, though chastised in the Hutton Report into the death of David Kelly, Andrew Gilligan's revelations about the weapons inspector brought the deliberate misleading of the British public over Iraq's nuclear capacity to the forefront of national attention. That said, this paternalistic model is undermined by Barthes. In his logic, the meaning of news reporting is inscribed by the audience, which undermines the integrity of journalistic standards. And indeed, over the past thirty years the proliferation of New Journalism styles and consumer discourse have blurred boundaries between the stylistic conventions of fiction and objective reportage. Moreover, the proliferation of lifestyle journalism has seen discrete routines of living become the focus of entire publications: from Angling Times to Living France, much professional journalism is orientated towards specific sub-cultural lifestyle groups. That the identification of these niche markets are of course a means of targeting defined communities of consumers by advertisers is of course documented in the work of Anderson. Indeed, with increased competition from the web-2.0 content in the shape of blogs, forums and message boards, traditional journalistic forms have had to become specialist products in their own right to justify the expense of purchasing a hard-copy of a magazine or newspaper. Hero brands like The Times, Guardian and Sun all have companion web site, while passion titles Kerrang and Q also exist as radio and television channels. Interactivity has become a key feature of traditional journalistic formats as online editions of articles invite the readers to solicit their own feedback, comments as well as engaging in inter-audience debate.

Conclusion

In reviewing three theories of consumer culture it is easy to see how they build upon ideas set out in the preceding chapter on postmodernism. In particular, underpinning all cultures of consumption is the belief that the purchase of material goods is intimately connected to the signification of identity. Most interesting perhaps, is the tension that exists between the way in which mass-produced goods are appropriated by mainstream and sub-cultural groups. That the improvisational meaning inscribed by alternatives sets are routinely used up and exploited by capitalist industries reinforces the neo-Marxist perspectives discussed in chapters 2 and 4. In this direction it would seem, however, that we are at a time of uncertainty: the proliferation of digital technology has made niche markets increasingly important to the way in which media texts are produced, and is impacting on fringe creative activities making them more economically viable.

9. Feminism

Introduction

In this chapter we look at the ways in which feminism offers resistive readings of texts, which challenge the patriarchal hegemony of Western thinking. Of particular significance will be how we theorise the post-feminist subject position. The chapter begins with an examination of Laura Mulvey's seminal article 'Visual Pleasure and Narrative Cinema' (1975). Though this is an over-cited work, it is a useful point of entry into the issues and debates relating to the politics of gender and contemporary media. The second section of the chapter turns to focus on the work of Helena Cixous; 'Sorties' (1975) was published at the same time as Mulvey's work but it offers a much more radical critique, drawing upon French feminists like Simone de Beauvoir and Julia Kristeva. Cixous advocates a rejection of the structuralist binaries and logocentric thinking, which she suggests invoke reductive definitions of gender and legitimise the subordinate position of women. Finally, in the concluding part of the chapter we embrace the work of Judith Butler, who argues that gender is performative and that there is no pre-discursive self that feminism must represent. Though it is over twenty years old, Butler's argument remains particularly persuasive as it aligns gender debates with the poststructuralist ideals of postmodernity. First, however, we turn to what is actually meant by the term feminism with specific reference to the work of Simone de Beauvoir.

9.1 Feminism – Simone de Beauvoir

NAME: Simone de Beauvoir (1908-1986)

KEY IDEA: De Beauvoir argues that the female subject is not born a woman, but rather she becomes one. Drawing upon Hegelian concepts of difference she contends that men make women 'other' so that they do not have to address their problems and in this sense women have much in common with other subaltern groups. Central to de Beauvoir conception of Feminism is the idea that women should not view masculinity as the norm to which they should aspire.

KEY TEXT: *The Second Sex* (1949).

Feminism is a contested term and can be used to refer to aspects of politics, culture, economics and sociology connected with protection and promotion of the rights, interests and welfare of women. In historical terms it can be defined in relation to the three specific epochs: first, second and third-wave feminism. Generally speaking first-wave feminism refers to the political activities of women at the end of the Nineteenth and beginning of the Twentieth Century. In Britain this is synonymous with the work of the suffragettes who campaigned for equal voting rights for women, which culminated in the 1918 Representation of the People Act. Central to this movement, of course, was the work of Emmeline Pankhurst who with the help of her daughters mobilised tens of thousands of women into active protest. Second-wave feminism is a term used to describe the development in the women's movement after the end of the Second World War. Unlike first-wave feminism, second-wave feminism organised itself in the form of scholarly writing. In particular the work of French writer and philosopher Simone de Beauvoir was very influential. In The Second Sex (1969) she argues that the female subject is not born a woman, but rather she learns to become one. 'Woman' in this sense is conceptualised as 'other': different and subordinate to masculine normality. Other French feminists of the post-war period that were particularly influential include Helena Cixous and Julia Kristeva. Kristeva's work, which emphasises the 'semiotic' pre-Oedipal stage of gender development is often labelled 'essentialist': believing in the universal characteristic of the feminine. By contrast, third-wave feminism is less interested in the idea of a fixed subject position that requires the kind of political agency that Emmeline Pankhurst fought for. Judith Butler is typical of third-wave feminists influenced by poststructuralist thinking for whom the gendered self is a contingent construct, signified and interpreted though a series of individual performances. In this sense, third-wave feminism has given rise to the term post-feminism to describe a generation of women for whom feminism has become a redundant concept. Academics like Angela McRobbie and Susan Faludi, however, remain sceptical of the term viewing it as populist conceit used by new right media to re-inscribe masculine hegemony. For the purposes of this chapter we focus on second and third-wave feminism because of their more explicit commentary on the realm of culture and the symbolic. In the concluding section consideration will be given to the way in which female celebrities like Madonna have come to embody the contradictions of a post-feminist subject position.

9.2 Laura Mulvey – The Male Gaze

NAME: Laura Mulvey (1941 – present)

KEY IDEA: Film theory can be informed by psychoanalysis. Mainstream Hollywood cinema puts the audience in a male subject position. By contrast, the female subject on screen is positioned in such a way that foregrounds physical attributes that will appeal to men. Mulvey calls this process 'the male gaze', which can be divided into two categories: the voyeuristic and the fetishistic. The former positions woman as sex objects while the latter places them on a pedestal to be worshipped.

KEY TEXT: *Visual Pleasure and Narrative Cinema* (1975).

Mulvey begins 'Visual Pleasure and Narrative Cinema' (1975) on the premise that she is appropriating psychoanalysis as a political weapon. However, she is dissatisfied with the way in which Freudian terms have been used by her contemporaries writing on cinema in Screen; in particular she takes issue with their phallocentric emphasis on the female's lack of a penis. She also emphasises the implications of this in semiological terms:

> Women then stands in patriarchal culture as signifier for the male other, bound by symbolic order in which man can live out his phantasies and obsessions through linguistic command by imposing them on the silent image of women still tied to her place as bearer of meaning, not maker of meaning. (Mulvey, 1975, 586)

On the cinema itself, she defines it as an advanced representation system devoted during the studio system to the creation of pleasure for its audience. With an eye to the future, however, she suggests this pleasure principle needs to be disrupted.

Invoking the work of Freud, Mulvey argues that cinema offers three possible sources of pleasure: the pleasure of being looked at; the pleasure of looking and the identification of what is taking place on the screen. Because cinema reflects a world of sexual imbalance Mulvey argues that pleasure in cinematic terms is coded in terms of gender: women are not only the object of the cinematic gaze but their knowing presentation of self communicates an awareness of being looked at. Reinforcing normative gender roles, Mulvey argues there is an 'active/passive' division of labour in films, with male characters tending to move the story along, while events 'happen to' female characters. For Mulvey, however, what makes film different to other medium is the controlled nature of the form:

101

Female subject on screen (The Male Gaze)

Playing on the tension between film as controlling the dimension of space (changes in distance, editing), cinematic codes create a gaze, a world, and an object thereby producing an illusion cut to the measure of desire (Mulvey, 595)

For Mulvey, the male gaze is just another aspect of the way in which the female form is stolen and manipulated in a patriarchal society and for this reason she contends that the decline of the studio system can only be viewed as a positive thing for women.

Mulvey's article focus's principally on mainstream Hollywood cinema of the 1940s and '50s, including the work of film auteur directors Alfred Hitchcock and Jonas Sternberg. In particular she talks about Sternberg's work with Marlene Dietrich:

Sternberg produces the ultimate fetish, taking it to the point where the powerful look of the male protagonist is broken in favour of the image in direct erotic rapport with the spectator. The beauty of the woman as object and the screen space coalesce; she is no longer the bearer of guilt but a perfect product whose body stylized and fragmented by close-up is the content of the film and the direct recipient of the spectator's look. (Mulvey, 592)

Hollywood stars like Marlene Dietrich and of course Marilyn Monroe are of course problematic for feminists because they embody this very masculine fantasy of the feminine. However, as Mulvey suggests their self-consciousness of being looked at offers an agency of sorts, albeit a highly restricted one. And, indeed though mainstream Hollywood films of the 1950s like George Sidney's Annie Get Your Gun (1950) and David Butler's Calamity Jane (1953) depict strong women, the object of both plots is to 'get the man'.

By contrast the 1970s was characterised by an upsurge in independent women filmmakers and films that explore feminist themes. Titles released during this period included Joyce Chopra's Joyce at 34 (1972), Margaret Lazarus's Taking Our Bodies Back (1974) and Amalie R. Rothschild's Nana, Mom and Me (1975): all depict women taking control of their own lives. In the proceeding years films by female directors have found great success including Fannie Flagg's Fried Green Tomatoes (1991), Jane Campion's The Piano (1993), Gurinder Chandra's Bend it Like Beckham (2002). However, in 2009 it is estimated that only 6% of film-makers are women and many of those films perceived by audiences to explore feminist themes are in fact directed by men, for example Steven Spielberg's adaptation of Alice Walker's novel The Colour Purple and Quentin Tarantino's epic Kill Bill (2003/4). Indeed, for all the revolutionary potential of the 1970s, contemporary Hollywood does not seem to have moved on significantly since the 1950s.

Arguably this stasis in the representation of women in mainstream Hollywood is a reflection of the broader complexities of feminism in the 1980s. This period saw the rise of the new right in Britain and America, under the leadership of Margaret Thatcher and Ronald Reagan. In particular, the introduction of social policy that reinforced the primacy of the family and the role of women as mothers challenged the advances made by the women's movement in the 1970s. And, indeed, for many theorists, like Susan Faludi, this was part of widespread backlash against feminism in the West during this period. The emergence then of a more ambiguous attitude towards the politics of gender in the 1990s and beyond can then be interpreted in different ways. For example, the proliferation of men's lifestyle titles like FHM and Loaded could be viewed as a re-inscription of a more singular notion of patriarchy in which the objectification of the female form is central. By the same token, the commoditisation of men's lifestyles during this period could be said to reflect a crisis in the definition of modern masculinity. If as Janice Winship has suggested, the need for women's magazines was a symbol of their exclusion from mainstream culture, the emergence of a parallel men's market speaks of it's own predicament. And of course the emergence of more salacious forms of tabloid press aimed at women has arguable introduced a female gaze in which both male and female forms are scrutinised and dissected. In this sense Mulvey's concept of the male–gaze can be just as usefully applied to areas dislocated from both cinema and gender.

In the 1990s Andrew Higson wrote about the aspirational gaze invoked by cinematography in films by the English production company Merchant Ivory. In particular he viewed the objectification of heritage as a way in which audience's configured an idealised vision of Britain's collective past as a commoditised form for private consumption. That same process of aspiration can be seen in the new media landscape of the Twenty-First Century: Facebook, MySpace and Bebo invite us to present idealised versions of ourselves. A self that is highly stylised and contrived. In consideration of Sternberg's work with Dietrich, Mulvey states that the actress is a 'perfect product'; that same process of objectification can be observed in the way in which our virtual reality selves are constructed. What this says about contemporary notions of gender is open to interpretation: however, it would seem, as Anthony Giddens suggests in the Transformation of Intimacy (1992) the personal sphere is inextricable from commoditised cultural forms.

9.3 Helena Cixous – Sorties

NAME: Helena Cixous (1937 – present)

KEY IDEA: Cixous challenges the binaries that underpin structuralist systems of knowledge: activity/passivity, sun/moon, nature/art. For Cixous these are reductive and a product of the same binary that positions women in a subordinate position to men. She argues that future conceptions of gender should be more fluid and that bisexuality offers a vision of complete being. Bisexuality she argues can be viewed in two ways: as the identification of two beings within one and an internal space in which there is no such difference.

KEY TEXT: 'Sorties' in *The Newly Born Woman* (1975).

Cixous's 'Sorties' begins with a rhetorical question 'Where is she?', which precedes a list of binary opposites: 'activity/passivity, sun/moon, culture/nature, day/night, father/mother, head/heart, intelligible/palpable, logos/pathos' (Cixous, 1975, 578). The answer is of course resounding: in the subaltern second category. Having drawn attention to this Cixous then proceeds to challenge the logocentricity of structuralist couplings. For Cixous such pairing is inevitably hierarchical and invokes the primary philosophical opposition between man and women: activity versus passivity. Either a woman does not exist and she is passive: the alternative is unthinkable.

In the first instance Cixous implores that we challenge the connection between logocentricism and phallocentricism. This, however, is problematic: the phallocentric order of society and culture under which the feminine is buried is weighed down by the credence of logocentric order. They are inextricably linked and a rejection of the former would require a complete overhaul of the latter. As Cixous states: 'all the history, all the stories would be there to retell differently; the future would be incalculable' (Cixous, 580). And in this sense the system of difference with which she began her article would all be different.

Though Cixous emphasises the strength of phallocentric order, she is optimistic about instances of resistance against it. In particular, she emphasises the importance of those who do not repress the homosexual element: creative thinkers according to Cixous are stirred by anomalies. She is not suggesting that you have to be homosexual to create but rather open to the possibility:

> There is no invention of any other I, no poetry, no fiction without a certain homosexuality (the I/play of bisexuality) acting as a crystallization of my ultrasubjectivities. I is this exuberant, gay, personal matter, masculine, feminine or other where I enchants. (Cixous, 581)

Two beings within (Sorties)

Bisexuality she argues can be viewed in two ways: as the identification of two beings within one and an internal space in which there is no such difference. In this sense Cixous acknowledges bisexuality has far more to offer the woman than the man; however, she takes things one step further by suggesting that all women are to varying degrees bisexual: women's writing is viewed as very important by Cixous because it reveals the other. The act of writing is in this sense a way in which the feminine brings itself into existence. Paradoxically, therefore, it is the very passive position of women (her capacity to de-appropriate herself) that enables this creative and plural other to come into full fruition: a culmination without end that is neither masculine nor feminine but human.

Cixous' emphasis on writing is typical of French feminism in the 1970s; Julia Kristeva and Xavier Gauthier also explore the relationship between women and language. The term 'ecriture feminine' is often used to describe the kind of writing that enabled the plural creative other to come into fruition and indeed the term was first used by Cixous in her essay The Laugh of the Medusa (1975). An example of a text that is widely considered to be a form of 'ecriture feminine' is The Yellow Wallpaper (1892), a short story by the American novelist Charlotte Perkins Gilman. In a sense The Yellow Wallpaper can be viewed as an allegory for Cixous's emphasis on writing. The central theme of the book is that the restricted way in which society views women, is detrimental to their mental and emotional wellbeing. Of particular pertinence to Cixous argument is not only the importance of writing to the narrators identity but also the way in which male characters in the story caution her against this, echoing the threat Cixous perceives female subjectivity poses to patriarchal order.

Ecriture feminine, however, is not without its problems. A key issue for feminist critiques is the patriarchal structure of language. Indeed, both Kristeva and Cixous consider the contradictions endemic to the appropriation of a phalli-centric sign system by a previous generation of feminists. On the one hand, women's writing is viewed as a separate symbolic form. On the other hand, as long as female writing is dependent upon male forms of expression it is in dialogue with those exclusionary practices. By the same token, the emphasis placed upon the oneness of the psyche and text, as well as the anti-theoretical stance of its proponents, renders 'ecriture feminine' at odds with the 'Foucauldian' notions of truth at its centre. Likewise, the focus upon the body is controversial. While in part the reproductive process is seen as an agent of patriarchal oppression, simultaneously, motherhood is presented as the fount of all female sensuality and something to be reclaimed. Similarly, disruption is viewed as commensurate with patriarchal notions that femininity is always synonymous with unreason and discord. Yet, without the creation of disorder it is impossible to see how those on the margins might resist oppression.

In the proceeding years, various female writers have experimented with the form of literary fiction. Most well known is the work of Kathy Acker, whose novels deliberately disrupt traditional narrative forms; parodying the style of other writers and in some cases borrowing and re-working passages from texts in a manner more common to sample based dance music than literature. Blood Guts in High School, Acker's most well known work was written in the late 1970s: the novel disrupts traditional narrative form and is best understood as a collage of drawings, poems, maps, dialogue and third-person narration. Like much of Acker's work it has upset some feminist critics who feel the explicit sexual nature of the book, which also has a strong theme of violence, degrades women and reinforces a patriarchal viewpoint. However, it is without doubt that her writing embodies the spirit of chaos and disorder that for some characterises ecriture feminine. More palatable perhaps is the work of Jeanette Winterson, whose first novel Oranges Are Not the Only Fruit (1985) met with considerable critical acclaim upon publication. The semi-autobiographical, coming of age story, which details the life of a girl growing up in the North of England and her realisation that she is a lesbian was adapted into a television series by the BBC in 1990. However, it was perhaps her subsequent novel 'Sexing the Cherry' (1989) that broke more ground for women's writing. Widely regarded as postmodern novel the story is set partly in 17th Century London and partly in the present; the protagonist and her mother journey across space and time to find 'the self' and the narrative features elements of intertextuality. Ultimately, however, like Acker, for all the sophistication in Winterson's manipulation of text level form, it is very difficult to over-come the need to write at word and sentence level in English.

9.4 Judith Butler – Gender Trouble

NAME: Judith Butler (1953- present)

KEY IDEA: There is no gendered subject position that exists prior to the performance of gender. Gender is in this sense a contingent event: the gendered subject is something to be accomplished and something that changes according to context. This challenges the central premise of Feminism that there is a political subject that requires agency.

KEY TEXT: Gender Trouble (1990)

In Gender Trouble, Judith Butler tackles the work of preceding second-wave feminism and argues that there is no ontological self that is feminine. The problem with this for Butler is that is limits the kind of experiences that can be articulated as part of feminist discourse. The challenge for feminism in Butler's logic therefore is reconciling the historic exclusion of women from hegemonic discourse, without limiting the inclusiveness of feminism. To put things more simply, feminism needs to recognise that the catch-all use of the term women conceals experiential differences linked to gender, race, ethnicity and social class:

> Clearly the category of women is internally fragmented by class, colour, age, and ethnic lines, to name but a few, in this sense honouring the diversity of the category and insisting upon its definitional nonclosure appears to be a necessary safeguard against substituting a reification of women's experience for the diversity that exists. (Butler, 1990, 327)

Like Cixous she views the term woman as a masculine construct that labels the feminine as other: different. However, instead of focusing on the primacy of writing she proposes that the self is constructed through a series of performances: stylised acts that may rely upon restricted frameworks to make semantic meaning but are arbitrary none the less.

The performance of gender (Gender Trouble)

In her emphasis upon disciplinary techniques and regulative discourse, Butler is of course invoking the work of Foucault on the function of bodily discipline in the organisation of human behaviour. Bentham's model of the Panopticon discussed in chapter 1 is exemplary of this. In Butler's logic, however, these regulative discourses help maintain the illusion of gender stability. Drawing upon the work of Julia Kristeva she suggests that in the pre-Oedipal 'semiotic' stage of development 'unconscious fantasies exceed the legitimating bounds of paternally organised culture' (Butler, 333). However, where she differs from Kristeva is in her disavowal of self that exists at all and in this sense Butler is rejecting psychoanalytical frameworks that categorise second-wave feminism.

Of course one argument that can be leveled at Butler is that she does not account for the material biological differences between the sexes. While she acknowledges that this exists, for her this is immaterial to the construction of gender, which she views as entirely cultural. In this sense Butler views 'sex' as an alibi, which naturalises the attribution of gender at birth. Overall Butler's view of gender is transforming because it aligns feminism with a more postmodern cultural paradigm. Indeed, in her reading of the parodic nature of gender her thoughts echo those of Baudrillard and Jameson on the simulacrum:

> The notion of gender parody defended here does not assume that there is an original which such parodic identities imitate. Indeed, the parody is of the very notion of an original. (Butler, 338)

In this sense Gender Trouble is definitive in understanding contemporary ideas about gender because at its core is a rejection of the distinction of the real and the simulated. Unlike preceding definitions of feminism, Butler's work is significant also because it speaks for the subject position of men as well women; while patriarchal oppression may be a culturally constructed regulative discourse, Butler's work acknowledges that it does not universally benefit the lives of all men.

One area in which Butler's work has been particularly influential is that of Popular Music Studies. In the wake of Gender Trouble, a number of feminists have explored the way in which popular music performance embodies Butler's ideas. In particular the work of the American singer Madonna has come in for some close scrutiny. Alice Kaplan, Cathy Schwichtenberg and Beverley Skeggs, for example, all wrote pieces in 1993 that debate the extent to which Madonna can be seen to exemplify Butler's arguments. Indeed, David Gauntlett uses the term 'Madonna Studies' to categorise analysis of the singer (www.theory.org). Influenced by Michel Foucault's History of Sexuality (1976) as well as Butler, such readings have the tendency to see musical performance as representative of radical gender politics. For example, in 'Guilty Pleasures Feminist Camp from Mae West to Madonna' (1996) Pamela Robertson asks the question 'Is Madonna a glamorized fuckdoll or the queen of parodic critique?' (Robertson, 1996, 188). The problem with 'Madonna Studies' from the perspective of Musicology is that very little analysis is focused on the musical text but rather performances and promotional video. In addition to this, in its application of Butler's work, 'Madonna Studies' has the tendency to conflate self-conscious and highly stylised musical performances with the more routinised performance of self. That said, within Popular Music Studies, there is some very forward thinking analysis of masculinity. In Sheila Whitely's Sexing the Groove (1997), for example, both Gareth Palmer and Stan Hawkins offer salient accounts of the way in which both Bruce Springsteen and the Pet Shop Boys manipulate their own gendered subject position. And in this sense it is possible to see the way in which ideas about the self and the gendered subject are framed and shaped by media stars.

The notion that the self is multiple is embedded in the history of British popular music. In part this can be attributed to the subaltern position of British pop in the 1950s in relation to Afro-American cultural forms: the likes of Tommy Steele and Cliff Richard always did invoke a parodic sensibility. And, in this sense it is possible to frame British popular music as what Susan Sontag (1964) would define as a specifically 'camp' semiotic event. That this is connotative also of some very specific codes of masculine sexual identity has been remarked upon by a number of critiques. As Richard Dyer (1990) suggests, British popular music has always acknowledged the sign of 'gayness'. However, it can also be seen as symbol of what Bakhtin categorises as the carnivalesque and the inversion of normative structures of power. And, indeed, British popular music is littered with example of what Bracewell (2006) defines as 'hetero-camp': nominally heterosexual stars who embrace the mutability and glitter normally associated with less singular constructions of self (David Bowie, Bryan Ferry, Marc Bolan etc). Without wishing to re-inscribe normative patriarchal structure, it could be argued, therefore, that it is in relation to the masculine pop subject that the application of Butler's thesis has the most to offer: not because that subject position is of higher status but because in the celebration of its own contrived performance, the masculine pop subject embodies Butler's own aspirations for contemporary masculinity.

Conclusion

In reviewing three feminist perspectives it is easy to see how they can shape and inform our understanding of the way in which contemporary texts make meaning. Most interesting perhaps, is the tension that exists between those feminists who view women as separate ontological beings from men, and those who view gender as an artificial cultural performance. In part this is because the debate is the backdrop to contemporary ideas about the role of women in a post-feminist age. Likewise, the fine line between the need to represent the political needs of women, and being reductive in our understanding of that subject position, has direct relevance in a society in which women perform multiple and sometimes conflicting roles simultaneously. Ironically however, it is perhaps the notion that gender does not exist that has most to offer women in the long run because it simultaneously calls into question both notions of the feminine and the taken-for-granted supremacy of patriarchal order.

10. Post-Colonialism

Introduction

In this chapter we look at the way in which post-colonial theories can be used to frame and shape the way in which we think about contemporary media, society and culture. The chapter begins with an overview of Edward Said's model of the sign system and way in which Western cultural perspectives objectify non-Western forms as exotic and other. The second section of the chapter turns to focus on the work of Paul Gilroy, his view of Diaspora and the way in which this bi-passes Western forms of cultural interpretation influenced by the Enlightenment. Finally, in the concluding part of the chapter we turn to the work of Homi Bhabha in a consideration of the nature and purpose of cultural hybridity in the representation of contemporary ideas about diaspora. First, however, we consider what is actually meant by the term post-colonialism.

Post-colonialism

Post-colonialism refers to a complex and competing set of discourses that consider the legacy and intellectual ramifications of colonialism. By colonialism we are of course talking about the process of colonisation intrinsic to Empire building: one country's claim to sovereignty over another. When referring to colonialism there is therefore a tendency to make implicit reference to the British Empire, which by the early Twentieth Century claimed jurisdiction over a quarter of the Earth's surface. Colonies of the British Empire included Canada, Australia, India, large parts of Africa and the Middle East. Other notable Empires, however, include French, Spanish and Dutch projects. While the latter half of the Twentieth Century saw the dissolution of most imperial forms of international associations, the legacy of unequal power relations between colonial rulers and indigenous population has influenced global relations into the Twenty-First Century. Laid over the complexity of political affiliations and international allegiances, post-colonialism is intimately connected to the politics of race embedded in that history including but not restricted to the slave trade, apartheid, and immigration.

From a European perspective, one of the ways in which the legacy of Empire and politics of post-colonialism is understood is the notion of multi-culturalism. In Britain, for example, the 1950s and '60s saw a proliferation of immigration from former colonies in India and the Caribbean: people who believed they would be welcome as British subjects in the motherland. That many immigrants to Britain faced a less than enthusiastic reaction from the indigenous white population is well documented. Sixty years on from the arrival of the S.S. Empire Windrush, though British society is more accepting of its black population and there are many examples of the way in which the Britain celebrates its cultural diversity. However, in some areas, ethnic groups are still very ghettoised. Burnley in Lancashire, for example, has a large population of second and third generations Pakistanis, who have not only had to deal with the social problems associated with unemployment, poor quality housing and healthcare but racially motivated violence from other white British inhabitants.

Contemporary perspectives on issues of race are in this sense often very selective in the way history is presented, as the rise in support for the British National Party is testimony too. What individuals like Nick Griffin forget is that the wealth of the West is in many ways based upon the historical exploitation of developing world over which we once ruled. In addition to this, in the case of Britain, our sovereignty as a nation in the Twenty-First Century owes a debt to those soldiers from the Empire that fought alongside white Britons to protect the UK from the rise of fascism in Europe. Furthermore, while the last fifty years has seen a rise in immigration from overseas, the same period has also seen large numbers of British people move abroad themselves to Australia, Canada, South Africa and most recently Mediterranean resorts in Spain, Italy and former Eastern block countries.

10.1 Edward Said – Orientalism

NAME: Edward Said (1935 – 2003)

KEY IDEA: Orientalism refers to the academic study of the Orient, however, for Said the term describes the tendency in Western intellectual and artistic discourse to view the Orient as "other" – an exotic outsider to the Occident. Moreover, he suggests that despite being shaped by the colonial expansion of the Nineteenth Century, 'Orientalism' is not an explicit mode of political power and repression but instead exists in the "them-and-us" exchange within 'various kinds of power'

KEY TEXT: *Orientalism* (1978).

Edward Said was a Palestinian-American literary theorist who was influenced by poststructuralist thinkers like Foucault and Derrida in considering the relationship between power, knowledge and cultural representations. In particular Said draws upon Foucault's notion of discourse and the embedded power relations described in The Archaeology of Knowledge and Discipline and Punish. He argues that to examine 'Orientalism as a discourse' is key to understanding European culture's ability to 'manage and even produce - the Orient' (Said, 1978, 873). Orientalism refers to the academic study of the Orient, however, for Said the term describes the tendency in Western intellectual and artistic discourse to view the Orient as "other" – an exotic outsider to the Occident. Moreover, he suggests that despite being shaped by the colonial expansion of the Nineteenth Century, 'Orientalism' is not an explicit mode of political power and repression but instead exists in the "them-and-us" exchange within 'various kinds of power':

> [P]ower political (as with colonial or imperial establishment), power intellectual (as with reigning sciences like comparative linguistics or anatomy, or any of the modern political sciences), power cultural (as with the orthodoxies and canons of taste, texts, values), power moral (as with ideas about what "we" do and what "they" do) (Said, 1978, 874).

Further to this, Said suggests he is not looking for something hidden or encoded within Orientalist texts, as a structuralist would, but instead seeks to analyse 'the text's surface, its exteriority to what it describes' (Said, 875). In other words he is suggesting that the discourse of Orientalism is so embedded in Western thought that it remains unchallenged - protected by an armour of accepted norms, values, representations and stereotypes.

One example of an explicit critique of Orientalist discourse in the media is in the music of Mathangi Arulpragasam or M.I.A. the Sri-Lankan born British hip-hop artist. M.I.A. uses her music and diasporic position to comment on the cultural division between East and West; in particular the polarisation between Christian and Islamic nations that has occurred as a result of America and Britain's military offensive on Iraq and Afghanistan known as the 'War on Terror'. Much of her work explores the stereotyping of 'the exotic other' embedded in contemporary British multiculturalism: the tourist-like sensibility through which Eastern culture is fetishised in the Western Metropolis. The song "Paper Planes" for example explores the potent mix of fear and fascination with which the Westerner exoticises an Asian taxi driver as a potential mugger or terrorist.

The exotic outsider (Orientalism)

M.I.A. exposes this view as Orientalist "othering" by juxtaposing such narratives with the mundane reality of a migrant worker living at subsistence level and therefore only concerned with taking the customers money in order to put food on the table! Paper Planes features on the soundtrack to the film Slumdog Millionaire (2009). The film both explores and explodes cultural stereotypes by bridging the gap between Orientalist fantasy (the seductive fairytale representations of India seen in the Bollywood movie), and economic reality (the subordination of an Indian workforce to the West and the huge disparity between rich and poor in the film's setting of Mumbai).

10.2 Paul Gilroy – The Black Atlantic: Modernity and Double Consciousness

NAME: Paul Gilroy (1956 – present)

KEY IDEA: There are two distinct ideological codes that apply to the cultures of Diaspora: 'the politics of fulfilment' (the ethical): and 'the politics of transfiguration' (the aesthetic), which is grounded in the experiential practices of the mimetic, dramatic, and performative. The key idea here is that both of these elements ignore the social-scientific rationalism of the Enlightenment; in effect bypassing the Modernist project and instead privileging practices that are in turn, socially situated and environmentally contingent; in short, the pre and postmodern.

KEY TEXT: *The Black Atlantic* (1993)

The cultural theorist Paul Gilroy offers a model of resistance to the Western tradition of Orientalist 'othering'. Gilroy argues that the Diaspora culture formed as a result of triangular relationship between Africa, American and Britain, and instigated by the transatlantic slave trade, has generated its own tradition, which is simultaneously utopian and idealist, postmodern and pragmatic. For Gilroy, transatlantic Diaspora culture, particularly as expressed in hybridised forms of popular music, offers a tradition of the de-traditional. It is imbued with a Marxist utopianism; a discourse that transcends the capitalist logic of modernity in its existentialist commitment to that which he terms 'the politics of fulfilment: the notion that a future society will be able to realize the social and political promise that present society has left unaccomplished in the face of capitalist oppression (Gilroy, 1993, 973). This is endemic in the codified forms of musical expression originating from the plantations, which evolved to supply the 'courage required to go on living in the present' of that 'racial terror' (Gilroy, 1993, 973).

115

Gilroy also suggests that whilst the culture of transatlantic Diaspora is one that consecrates the present, equally it almost transcends 'modernity, constructing both an imaginary anti-modern past and a postmodern yet-to-come' (Gilroy, 1993, 974). So in effect he is suggesting that there are two distinct ideological codes that apply to the cultures of Diaspora. On the one hand there is 'the politics of fulfillment (the ethical), which refers to a discourse of utopianism firmly rooted in the 'semiotic, verbal, and textual' practices of everyday life; and on the other hand there is 'the politics of transfiguration' (the aesthetic), which is grounded in the experiential practices of 'the mimetic, dramatic, and performative' (Gilroy, 1993, 974-975). The key idea here is that both of these elements ignore the social-scientific rationalism of the Enlightenment; in effect bypassing the modernist project and instead privileging practices that are in turn, socially situated and environmentally contingent; in short, the pre and postmodern. In light of this assessment of Diaspora culture Gilroy proposes the following:

> ...that we reread and rethink this expressive counterculture not simply as a succession of literary tropes and genres but as a philosophical discourse which refuses the modern, occidental separation of ethics and aesthetics, culture and politics (Gilroy, 1993, 975).

Gilroy's book is enlightening not only because it provides clues as to how we might conceive multiculturalism in an anti-essentialist and organic way but also as it suggests that postmodern discourse has been misconceived as somehow estranged from that which has come before it. However, for Gilroy postmodernism is connected to a pre-Enlightenment notion of communality. If we expand his analysis of transatlantic Diaspora music to incorporate contemporary music distribution and consumption then it is easy to see how 'The Black Atlantic' is a useful model for interpreting youth culture in the age of the Internet. Whilst a more obvious linage can be found in the rave culture of the Nineties, in Internet music culture the possibilities for cross-cultural fertilisation are endless. Moreover social-networking sites, Internet forums and blogs are transforming geographically specific niche genres into popular music phenomena. One such hybrid genre that has made this transition is Grime.

Grime is a hybrid musical form, which is a product of the Diaspora cultures of East London, specifically Hackney and Tower Hamlets. Grime is a hybrid of American hip-hop, Jamaican dancehall and a continuum of the break-beat music of Britain's hardcore (rave) and drum 'n' bass scenes. In other words it is a reflection of the shifting cross-cultural relations of transatlantic Diaspora. The sound of grime is both edgy and aspirational, in many ways less about a consistent sound – which is constantly evolving – than an attitude; one of ambition, emancipation and redemptive 'transfiguration'. This transitional dimension is explicitly audible in the music of grime's most commercially successful artist Dizzee Rascal, whose sound has evolved from a dark sonic representation of the urban outsider to a playful pallet of cartoon-like hyper real sounds.

The politics of transfiguration (The Black Atlantic)

The dynamism of grime arguable stems from two main factors; firstly there is the obvious need to rise out of the hopelessness of urban deprivation; to be reborn in the world of privilege to which a large part of British society has come to take for granted. Secondly, the urgency of the music may stem from the obvious difficulty for black British artists to become commercially successful. Less than 2% of the UK's population is black compared with 13% in America. This ultimately means that sales of black music are relatively low in Britain and therefore a black artist has to become popular in the US in order to be commercially successful. This factor has arguably sustained the dialogue of transatlantic Diaspora; and has therefore called for the genre to re-articulate the forms and conventions of US black music, specifically those of "gangsta" rap and hip-hop.

10.3 Homi Bhabha – The Location of Culture

NAME: Homi Bhabha (1949 – present)

KEY IDEA: Bhabha's central argument is that migration has led to a cultural 'hybridity', which destabilises the traditional narrative tendencies of Western thought. Key to this idea is the ability of migrant peoples to re-envision the world from an external perspective, looking in on traditions rather than looking out from within them. Central to this is the notion of renewal, and the ability of Diaspora cultures to breath new life into old ways.

KEY TEXT: The Location of Culture (1993)

In The Location of Culture (1994) Bhabha set outs a poststructuralist agenda to undermine the system of binary oppositions through which Western philosophy, sciences and indeed the arts, have defined the cultural "other" as outsider. He has achieved this by relocating the philosophical centre away from the grand narratives of metaphysics and firmly in the existential Heideggerian realm of what he terms the 'performative' present. Bhabha suggests that in post-colonial society, culture is never on one side or the other of a binary division but instead on the 'boundary'. The historical narrative of 'homogenous national cultures' is broken by the story of migration – that of the politically and geographically displaced diaspora (Bhabha, 1994, 936). Furthermore he argues that post-colonialism does not simply represent the separation of sovereign state and colony but their ongoing relationship 'within the "new" world order and the multinational division of labour' (Bhabha, 1994, 937).

118

Call centre (The Location of Culture)

Bhabha's central argument is that migration has led to a cultural 'hybridity', which destabilises the traditional narrative tendencies of Western thought. He illustrates this through the analysis of works by Diaspora writers, suggesting that the novel enables us to see 'inwardness from the outside' (Bernasconi, 1991, 90). Key to this idea is the ability of migrant peoples to re-envision the world from an external perspective, looking in on traditions rather than looking out from within them. Central to this is the notion of renewal, and the ability of Diaspora cultures to breath new life into old ways:

> The borderline work of culture demands an encounter with "newness" that is not part of the continuum of past and present. It creates a sense of the new as an insurgent act of cultural translation. Such art does not merely recall the past as social cause or aesthetic precedent; it renews the past, refiguring it as a contingent "in-between" space, that innovates and interrupts the performance of the present. The "past-present" [binary] becomes part of the necessity, not the nostalgia, of living (Bhabha, 1994, 938).

The British film director Gurinder Chadha and writer Meera Syal explore the relationship between tradition and the 'performance' of cultural hybridity in Bhaji on the Beach (1993). In particular the film illustrates Bhabha's notion of the performative of 'mimicry'; that for hybrid cultures to integrate they must pastiche tradition and reconstruct it 'around an ambivalence'; in other words to produce simulacra such as the hyper-real environment of the Indian restaurant. Furthermore he argues that for mimicry to be effective it 'must continually produce its slippage, its excess, its difference' (Bhabha, 1994, 86). This idea is emphasised by the setting for the film, Blackpool, which exemplifies the carnivalesque aesthetic of difference. Blackpool is a global theme park; every culture is another ride and must be performed in its own distinct setting. This is reinforced by the disruption caused by one of the lead characters when she tries to eat her own "Indian" food in a seaside café; to which she is told: 'If it's takeaway you want, the "Khyber Pass" is around the corner'.

The correlation between the plurality of the black subject and the popular culture of hybridisation in Bhaji on the Beach is perhaps dependant on the film's depiction of Blackpool. However, in so far as Blackpool draws parallels between American cultural imperialism and Asian Diaspora, it is also used to make some clear distinctions. On the one hand Chanda seems to be saying that there are elements of Diaspora integral to British culture, and this is certainly implicit in some of dialogue between the characters of Ambrose Waddington and Asha: 'There used to be eleven live venues before the war… Opera, royal premiere classics. That was our popular culture then. Look at what we've become'. On the other hand, there is a sense that the black subject is fetishised in today's pluralistic consumer society. As is exemplified by Julian Samuel's praise of Bhaji on the Beach as a 'Black British Masterpiece' it seems forever necessary to externalise the black subject from the mundane culture of everyday and to compartmentalise it in the realm of the carnivalesque or as Bhabha suggests to 'continually produce its excess' (Bhabha, 1994, 86).

10.3 Conclusion

In reviewing three texts that deal with the issues and debates surrounding post-colonialism we have hopefully highlighted the complexities of the issue. Most interesting perhaps, is the tendency to view non-Western cultural forms as other: the exoticisation of the unknown and the foreign. Such perspectives are in essence a form of cultural imperialism and reinforce the hegemony of the Western white male. By contrast, the concept of Diaspora challenges the taken-for-granted nature of those assumptions. Instead of viewing the contradiction of cultural hybridism as problematic, such ideological perspectives view these dialogues of difference as progressive: challenging the hegemony of Western patriarchal culture.

Bibliography

Adesioye, Lola. Invisible in plain sight in The Guardian Friday 5 September 2008. Retrieved from http://www.guardian.co.uk/music/2008/sep/05/urban.dizzeerascal

Adorno, Theodor. W. (1975). The Culture Industry Reconsidered in The Culture Industry: Selected Essays on Mass Culture by J.M. Bernstein London: Routledge, 1991.

Adorno, Theodor. W and George Simpson. (1942). On Popular Music in A Critical and Cultural Theory Reader by A. Easthope and K. McGowan. Milton Keynes: Open University Press, 1992.

Bakhtin, Mikhail. (1965). The Carnival and the Carnivalesque in Cultural Theory and Popular Culture by J. Storey. London: Prentice Hall, 1998.

Barthes, Roland. Music Image Text. New York: Hill and Wang, 1977.

Baudrillard, Jean. Seduction. London: MacMillan Press, 1979.

Baudrillard, Jean. (1981) Simulations in Continental Philosophy by Kearney, D M. Rasmussen. London: Blackwell, 2001.

Baudrillard, Jean. The Consumer Society. London: Sage Publications, 1970.

Bentham, Jeremy (1786). The Principles of Morals and Legislation. New York: Prometheus Books, 1988.

Bernasconi, Robert. Quoted in Levinas's Ethical Discourse in Re-Reading Levinas. Bloomington: Indiana University Press, 1991.

Bhabha, Homi. The Location of Culture. London: Routledge, 1994.

Bhabha, Homi (1994). The Location of Culture in Literary Theory: An Anthology by J. Rivkin. London: Blackwell Press, 1998.

Bourdieu, Pierre. Distinction: A Social Critique of The Judgment of Taste. London: Routledge, 1979.

Boykoff Maxwell T. & S. Ravi Rajan. Signals and noise: Mass-media coverage of climate change in the USA and the UK. EMBO reports Vol 8, No 3, 2007.

Brunsden, Charlotte and David Morley. Everyday Television: Nationwide in Popular Film and Television by T. Bennett. London: BFI Publishing, 1981.

Burke, Edmond. A Philosophical Inquiry into the Origin of Our Ideas of the Sublime and Beautiful. London: R. & J. Dodsley, 1757.

Butler, Judith. Contingent Foundations in Feminist Contentions: A Philosophical Exchange. By Seyla Benhabib. New York: Routledge, 1995.

Butler, Judith. Gender Trouble. London: Routledge.

Butler, Judith. Imitations and Gender Insubordination in Literary Theory: An Anthology by J. Rivkin and M. Ryan. London: Blackwell, 1998.

Cixous, Helene. 'Sorties' from The Newly Born Woman. University of Minnesota Press, 1975.

Corner, John. and Sylvia Harvey. Enterprise and Heritage: Crosscurrents of National Culture. London: Routledge, 1991.

Ewen, Stuart. All Consuming Images: The Politics of Style in Contemporary Culture in Literary Theory: An Anthology by J. Rivkin. London: Blackwell Press, 1988.

Faludi, Susan. Backlash: The Undeclared War Against Women. London: Vintage, 1992.

Featherstone, Mike. Consumer Culture and Postmodernity. London: Sage, 1991.

Foucault, Michel (1970). The Order of Things: An Archaeology of the Human Sciences in Literary Theory: An Anthology by J. Rivkin. London: Blackwell Press, 1998.

Foucault, Michel. The Birth of the Clinic: An Archaeology of Medical Perception. London: Tavistock, 1973).

Foucault, Michel. The Will To Knowledge. London: Penguin Books, 1976.

Foucault, Michel. Discipline and Punish: The Birth of the Prison. London: Penguin Books, 1977.

Giddens, Anthony. The Consequences of Modernity, London: Polity Press, 1990.

Giddens, Anthony. The Transformation of Intimacy. Cambridge: Polity Press, Cambridge, 1992.

Gilroy, Paul. The Whisper Wakes, The Shudder Plays in Race, Nation and Ethnic Absolutism: There Ain't No Black in The Union Jack. London: Routledge, 1987.

Gilroy, Paul (1993). The Black Atlantic. in Literary Theory: An Anthology by J. Rivkin. London: Blackwell Press, 1998.

Goffman, Erving. The Presentation of the Self in Everyday Life, (New York: Anchor Books, 1959.

Goodwin, Anthony. Sample and Hold – Pop Music in the Digital Age of Reproduction in On the Record – Rock, Pop and the Written Word by S Frith and A Goodwin. London: Routledge, 1990.

Gramsci, Antonio. The Antonio Gramsci Reader: Selected Writings 1916 to 1935. New York: NYU Press, 2000.

Hall, Stuart (1973). Encoding and Decoding in the Television Discourse in Culture, media, language. London: Hutchinson, 1989.

Hall, Stuart. Culture, Media, Language. London: Hutchinson Press, 1980.

Hall, Stuart. Notes on Deconstructing the Popular in Peoples History and Socialist Theory. London: Routledge, 1981.

Hebdige, Dick. Subculture: The Meaning of Style. London: Routledge, 1979.

Hebdige, Dick. The Bottom Line on Planet One: Squaring Up to The Face in Hiding in the Light. London: Routledge, 1988.

Heidegger, Martin (1962). Being and Time, Translated by John Macquarrie and Edward Robinson. Oxford: Blackwell, 1986.

Higson, Andrew. Re-presenting the National Past: Nostalgia and Pastiche in the Heritage Film in British Cinema and Thatcherism by L Friedman. London: UCL Press, 1993.

Jameson, Frederic. The Politics of Theory: Ideological Positions in the Postmodernism Debate in The Ideologies of Theory Essays 2. London: Routledge, 1988.

Jameson, Frederic. Postmodernism or The Cultural Logic of Late Capitalism. London: Verso Press, London, 1991.

Kant, Emmanuel. Answering the Question: What Is Enlightenment? In Berlin Monthly. Berlin: Berlin Monthly, 1784.

Kaplan, Ann. Rocking Around The Clock: Music Television, Postmodernism, & Consumer Culture. London: Methuen, 1987.

Kaplan, Ann. Whose Imaginary? The Televisual Apparatus, the Female Body and Textual Strategies in Select Rock Videos on MTV in Media Studies: A Reader by P. Marris and S. Thornham. Edinburgh: Edinburgh University Press, Edinburgh, 1996.

Kaplan, Ann. Madonna Politics: Perversion, Repression, or Subversion? Or Masks and/as Master-y in The Madonna Connection: Representational Politics, Subcultural Identities, and Cultural Theory by C. Schwichtenberg. Colorado: Westview Press, 1993.

Keightley's, Keir. Reconsidering Rock in The Cambridge Companion to Rock and Pop edited by S Frith W Straw and J Street. Cambridge: Cambridge University Press, 2001.

Klein, Naomi. No Logo, London: Flamingo, 2000.

Kristeva, Julia. Revolution in Poetic Language. Columbia: Columbia University Press, 1984.

Lacan, Jacques. The Mirror Stage in Ecrits. London: W.W. Norton and Co., 2007.

Lash, Scott (1994). Reflexivity and its Doubles: Structure, Aesthetics, Community, in Reflexive Modernization: Politics, Tradition and Aesthetics in the Modern Social Order by Ulrich Beck, Anthony Giddens and Scott Lash, Cambridge: Polity Press, 2004.

Levi Strauss, Claude. The Savage Mind. Chicago: University Of Chicago Press, 1963.

McRobbie, Angela. Postmodernism and Popular Culture. London: Routledge, 1994.

McRobbie, Angela. Settling Accounts with Subcultures: a Feminist Critique in Screen Education 34. Glasgow: University of Glasgow Press, 1980.

Mercer, Kobena. Welcome to the Jungle. London: Routledge. 1994.

Marcuse, Herbert. One Dimensional Man. London: Sphere, 1968.

Marwick, Arthur. British Society Since 1945. London: Penguin, 1996.

Marx, Karl and Friedrich Engels. (1848). The Communist Manifesto. London: Norton.

Middleton, Peter. The Inward Gaze Masculinity and Subjectivity In Modern Culture. London: Routledge, 1992.

Miller, Daniel. Consumption as the Vanguard of History. London: Routledge, 1993.

Morris, Meaghan. Banality in Cultural Studies in Literary Theory. London: Routledge, 1992.

Mort, Frank. Cultures of Consumption. London: Routledge, 1996.

Mort, Frank. The Commercial Domain: Advertising and the Cultural Management of Demand in Commercial Cultures by P Jackson. Oxford: Berg, 2000.

Mulvey, Laura Visual Pleasure and Narrative Cinema in Screen 16 (3), 1975.

Nixon, Sean. Hard Looks: Masculinity, Spectatorship and Contemporary Consumption. London: UCL Press, 1996.

Nixon, Sean. In Pursuit of the Professional Ideal: Advertising and the Construction of Commercial Expertise in Britain 1953 – 64 in Commercial Cultures by P Jackson. Oxford: Berg, 2000.

Packard, Vance. The Hidden Persuaders. London: Penguin Books, 1957.

Packard, Vance. The Waste Makers. London: Penguin Books, 1954.

Palmer, Gareth. Springsteen and Authentic Masculinity in Sexing The Groove – Popular Music and Gender by Sheila Whitely. London: Routledge, 1997.

Peirce, Charles Sanders. Logic as Semiotic: The Theory of Signs in Philosophical Writings of Peirce by Justus Buchler, New York: Dover Publications, 1966.

Radway, Janice. Reading The Romance. Carolina: University of North Carolina Press, 1984.

Railton, Diane. The Gendered Carnival of Pop in Popular Music 20 (3). Cambridge: Cambridge University Press.

Roberts, Kenneth. Contemporary Society and the Growth of Leisure. London: Longman, 1978.

Robertson, Pamela. Guilty Pleasures - Feminist Camp from Mae West to Madonna. London and New York: I. B. Taurus & Co., 1996.

Saussure, Ferdinand. Course in General Linguistics in Literary Theory: An Anthology by J. Rivkin. London: Blackwell Press, 1998.

Schwichtenberg, Cathy. Madonna's Postmodern Feminism: Bringing Margins to the Center in The Madonna Connection: Representational Politics, Subcultural Identities, and Cultural Theory by C Schwichtenberg. Colorado: Westview Press, 1993.

Segal, Lynne. Slow Motion: Changing Masculinities Changing Men. London: Virago Press, 1990.

Skeggs, Beverley. A Good Time For Women Only in Deconstructing Madonna by F Lloyd. London: Batsford, 1993.

Sontag, Susan. Notes on Camp in Camp: Queer Aesthetics and the performing subject: A Reader, edited by F. Cleto. Ann Arbor: University of Michigan Press. 53-65, 1964.

Storey, John. Cultural Theory and Popular Culture. London: Prentice Hall, 1994.

Taylor, Charles. Sources of Self, (Cambridge: Cambridge University Press, 1989.

Taylor, Paul. Investigating Culture and Identity. London: Harper Collins, 1997.

Thomas, Lynne. Fans, Feminisms and 'Quality' Media. London: Routledge, 2002.

Thompson, Grahame. The Carnival and the Calculable: Consumption and Play at Blackpool in Formations of Pleasure edited by Frederic Jameson and Terry Eagleton. London: Routledge, 1983.

Twitchell, James B. Romantic Horizons: aspects of the sublime in English poetry andpainting, 1770-1850. Columbia: University of Missouri Press, 1983.

Veblen, Thorstein. The Theory of the Leisure Class. New York: MacMillan, 1899.

Volosinov, Valentine. N. Marxism and the Philosophy of Language (1929) in Literary Theory: An Anthology edited by J. Rivkin. London: Blackwell Press, 1998.

Whitely, Sheila. Artifice and the Imperatives of Commercial Success – From Brit Pop to the Spice Girls in Women and Popular Music: Sexuality, Identity and Subjectivity. London: Routledge, 2000.

Whitely, Sheila. Sexing The Groove – Popular Music and Gender. London: Routledge, 1997.

Willis, Paul. A Theory for the Social Meaning of Pop in Stenciled Occasional Paper, Sub and Popular Culture Series: SP 13. Birmingham: Centre for Contemporary Cultural Studies, 1974.

Willis, Paul. Learning to Labour: How Working Class Kids Get Working Class Jobs, (Farnborough: Saxon House, 1977.

Winship, Janice. Inside Women's Magazines. London: Pandora Press, 1987.

Woods, Philip. Parents As Consumer Citizens in Ruling The Margins: Problematising Parental Involvement edited by R Merton, D Mayers, A Brown and J Vass. London: University of North London Press, 1993.

CPSIA information can be obtained at www.ICGtesting.com
Printed in the USA
LVOW09s0145050716

495092LV00030B/970/P

9 781523 940691